SUMMER STARGAZING

A PRACTICAL GUIDE
FOR RECREATIONAL ASTRONOMERS

TERENCE DICKINSON

FIREFLY BOOKS
BOOKMAKERS PRESS

Cataloguing-in-Publication Data

Dickinson, Terence
    Summer stargazing

Includes index.
ISBN 1-55209-014-0

1. Stars – Observers' manuals.
2. Astronomy – Amateurs' manuals.
I. Title.

QB63.D53 1996   520   C95-933246-4

A FIREFLY BOOK

Published by
Firefly Books Ltd.
3680 Victoria Park Avenue
Willowdale, Ontario
Canada M2H 3K1

Published in the U.S. by
Firefly Books (U.S.) Inc.
P.O. Box 1338, Ellicott Station
Buffalo, New York 14205

Produced by
Bookmakers Press Inc.
12 Pine Street
Kingston, Ontario
K7K 1W1

Design by
Roberta Cooke

Printed and bound in Canada
by Friesens
Altona, Manitoba

Printed on acid-free paper

This book is dedicated to my mentors, Helen, Ray, Henry and Ian.

Other Firefly books by Terence Dickinson
NightWatch
Exploring the Night Sky
Exploring the Sky by Day
The Universe and Beyond
The Backyard Astronomer's Guide
    (with Alan Dyer)
From the Big Bang to Planet X
Extraterrestrials (with Adolf Schaller)
Other Worlds

Front Cover:
The summer Milky Way with totally eclipsed Moon superimposed.
Photos by Terence Dickinson.
Cover design by Roberta Cooke.
Back Cover:
Partially eclipsed Moon and Lagoon Nebula region of the summer Milky Way (top); observer at dusk (bottom).
Photos by Terence Dickinson.

CONTENTS

Summer Stargazing

Stargazing is a personal exploration of the universe—an exhilarating, mind-expanding voyage in time and space

ABOUT THIS BOOK
Summer Stargazing is both a stargazing guide and a pictorial celebration of the summer night sky. Although primarily intended for use from May through September, it also contains guide charts for the entire year.

Imagine a summer evening far from the city. The air is refreshingly cool and still. As the sky darkens, stars begin to appear, and soon, a glittering tapestry has emerged in the blackness overhead. You sink back into a reclining lawn chair, and your mind reaches out to embrace the immensity before you. What unknown worlds float in that abyss? What civilizations have arisen in the galaxy's spiraling arms since the beginning of time? Where did it all come from?

It is no accident that the starry night stirs in us the most profound questions of origins, destinies and the ultimate meaning of it all. We are, in a very real sense, children of the stars. Every atom in our bodies and in the air we breathe was brewed billions of years ago in the fiery hearts of stars. All atoms except hydrogen and helium were (and are) by-products of the production of starlight. When stars die, either as bloated red giants or as explosive supernovas, vast cargoes of freshly cooked atoms spill into space, where they eventually collect in gas clouds called nebulas—the birthplaces of new stars. Our Sun and its planet family were born in just such a setting five billion years ago. Earth, air, rocks and people are all star-stuff.

THE LURE OF STARGAZING

After more than 35 years of avid skywatching, I have yet to lose the sense of awe I felt when I first stood under the stars as a child. If anything, the wonder and astonishment have been amplified as I have come to know the names, distances and enormous dimensions of the stars, planets and galaxies

on display overhead every night. That's the lure of backyard astronomy: *understanding what you are seeing.*

Once a few stars and constellations become readily recognizable, the night sky is transformed from a random array of glittering points to a

colossal mural of constellation patterns that moves in a stately seasonal rhythm. Upon that tapestry, you can follow the wanderings of the planets and the cycles of the Moon's phases. Or you can reach deeper and seek delicate star clusters and the wispy glow of star-forming nebulas. Then there's the unpredictable flash of a meteor, the occasional glowing auroral curtain and—beyond it all— the backbone of our galaxy: the Milky Way.

The universe unfolds in the night sky, and you can explore it in amazing detail with nothing more sophisticated than standard birdwatching binoculars. Of course, a decent telescope will reveal cosmic wonders inaccessible to binoculars, but for casual backyard or vacation viewing, binoculars are remarkably versatile. Using binoculars, I have seen the ghostly images of galaxies 30 million light-years from Earth. You can too.

GETTING STARTED

If you have ever tried stargazing, you already know that the most important initial hurdle is making the first, definite identification of a star, planet or constellation. Star charts can be intimidating, but there is a key to unlocking and relating them to the night sky. That key is the Big Dipper, visible every night of the year from the United States and Canada. This easily recognized configuration of seven stars is all you need to be on your way to a rewarding pastime.

SUMMER BRINGS *with it fine stargazing weather; it also happens to be the time of year when our galaxy, the Milky Way (photo below), arches high across the sky.*

CELESTIAL PHOTOGRAPHY vs. WHAT THE EYE SEES

Unlike human vision, time exposures on film accumulate light over many seconds, minutes or even hours to make faint objects bright and thus enhance subtle detail. Depending on the type of object being photographed, there can be significant differences between what the photo shows and what the eye would see. For example, this twilight view of the Moon and two planets (above) pretty much corresponds to the way I remember the scene when I photographed it. On the other hand, the wide-angle view of the summer sky below was a 20-minute exposure that makes the Milky Way a bit more prominent than it usually appears to the naked eye. Likewise, the aurora photo on facing page is more intense than it appeared to the eye (but not by much; it was a fantastic display). Extending this trend further, the cover photo and the Milky Way close-up at left, both photographed under superb mountaintop conditions, reveal significantly more color and detail than can be seen visually, either by the naked eye or with optical aid.

But before I get to the specifics of how to use the Big Dipper as a celestial signpost, some introductory information on the next few pages will help prepare you for the voyage.

Our Place *in the* Universe

Earth orbits a star that resides in an outlying spiral arm of an ordinary galaxy within a universe of 100 billion galaxies

If each star in the universe were a grain of sand, then all the stars visible on a dark, moonless night would fill a thimble. A large wheelbarrow would contain enough grains of sand to represent our galaxy, the Milky Way. But there isn't enough sand on all the beaches on Earth to represent the entire universe.

The immensity of the universe goes beyond these colossal numbers to encompass vast distances. Light, which travels 300,000 kilometers (186,000 mi) in one second, offers a perfect measuring stick. For example, the Moon is just over one light-second away. The giant planet Jupiter is half a light-hour distant, while Pluto,

at the rim of the Sun's far-flung planetary family, is five light-hours from Earth. Four light-years away is the nearest star, Alpha Centauri, marked in the illustration below.

Continuing outward, the Sun is 24,000 light-years from the star-rich hub of the Milky Way Galaxy. Our galaxy is, in turn, 2.3 million light-years

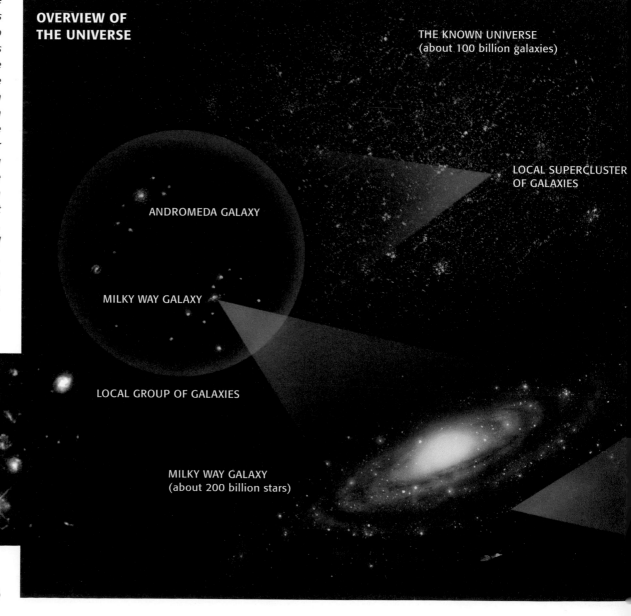

OVERVIEW OF THE UNIVERSE

THE KNOWN UNIVERSE (about 100 billion galaxies)

LOCAL SUPERCLUSTER OF GALAXIES

ANDROMEDA GALAXY

MILKY WAY GALAXY

LOCAL GROUP OF GALAXIES

MILKY WAY GALAXY (about 200 billion stars)

from the Andromeda Galaxy, our nearest major galactic neighbor. The most remote galaxies seen by the Hubble Space Telescope are roughly 10 billion light-years from Earth.

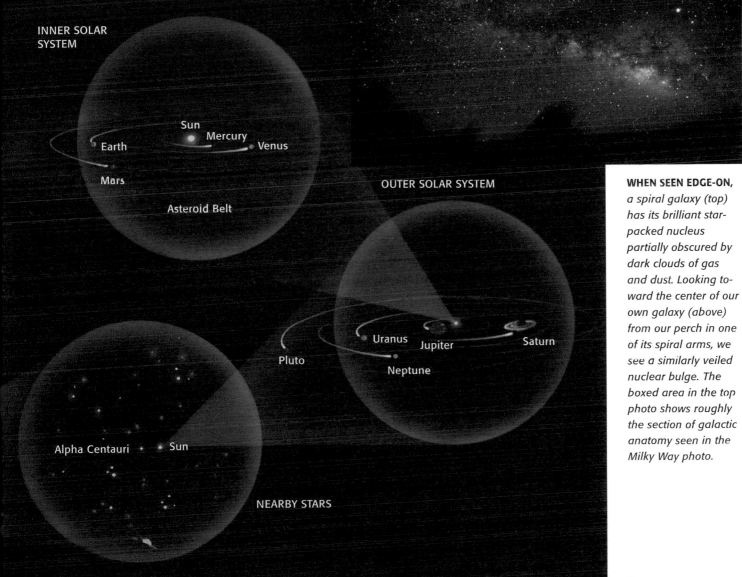

INNER SOLAR SYSTEM

Sun
Mercury
Earth
Venus
Mars
Asteroid Belt

OUTER SOLAR SYSTEM

Uranus
Jupiter
Saturn
Pluto
Neptune

Alpha Centauri
Sun

NEARBY STARS

WHEN SEEN EDGE-ON, *a spiral galaxy (top) has its brilliant star-packed nucleus partially obscured by dark clouds of gas and dust. Looking toward the center of our own galaxy (above) from our perch in one of its spiral arms, we see a similarly veiled nuclear bulge. The boxed area in the top photo shows roughly the section of galactic anatomy seen in the Milky Way photo.*

Stargazing Equipment

A few tips on what you need—and what to avoid—when starting out in astronomy

BINOCULARS
are the only essential piece of equipment you will need to enjoy recreational astronomy. A binocular-tripod adapter affixed to the glasses (above) is highly recommended. Eventually, you may graduate to a serious piece of astronomy hardware, such as this 8-inch Schmidt-Cassegrain telescope.

Unlike some other leisure activities, astronomy is not an "instant" hobby to which you can gain entry simply by purchasing fancy equipment. A telescope does not make its owner an amateur astronomer, nor is one necessary when learning the fundamentals. In fact, the fundamentals—learning your way around the night sky—should come first, before the telescope. A telescope is of little use if you don't know where to point it. An initial period of naked-eye sky observing complemented by binoculars is the route I recommend. This allows a gradual approach, like stepping into the shallow end of the pool rather than jumping into the deep end before you have learned how to swim.

But too often, this isn't what happens. Under the mistaken impression that a telescope is the first requirement, many budding astronomers buy one immediately. The result is usually disappointment, because the telescopes typically bought by (or for) beginners are those which are the most difficult to use. A standard $200 department-store or camera-store "astronomical" telescope—the type purchased by someone with good intentions but little practical knowledge of astronomy —invariably has a jiggly tripod and mount, plus inferior optical components. These instruments are frustrating to use and, in my opinion, not worth the expense.

Decent binoculars, on the other hand, are essential for both the casual observer and the veteran skywatcher. Standard bird-watching binoculars reveal more than 10 times the number of stars visible to the eyes alone, and the wide field

of view offered by all binoculars makes them much easier to aim than a telescope. Once you know your way around the sky, hundreds of subtle objects, such as nebulas (star-forming regions) nestled in the Milky Way and galaxies several million light-years from Earth, are possible viewing targets.

For recreational astronomy, the best sizes are 7x42, 8x40, 7x50 and 10x50. All binoculars have this two-number designation, the first being the magnification, or power, the second the diameter of the main lenses in millimeters. Sizes under 40mm don't collect enough light to reveal faint celestial objects, and glasses with main lenses over about 56mm are uncomfortably heavy. The binocular line-up at right clearly shows the substantial jump in size once the main lenses reach 60mm diameter. As for magnification, 7 to 10 power

8x20

screws into a threaded hole at the front of the binoculars' focusing bar. The base of the L attaches to the tripod like the base of a camera. When buying new binoculars, ensure that they have the threaded receptacle—not all do. If you already own binoculars without the threaded hole, check with

Another useful stargazing accessory is a red-filtered flashlight for consulting star charts at night without zapping the eyes' dark adaptation in the process. Use an ordinary penlight, and filter the illumination using layers of red cellophane or plastic affixed with elastics or tape.

READY FOR A TELESCOPE?

Suppose you feel you are ready to graduate to a telescope. What do I recommend? What counts is a rock-solid mounting assembly and quality optics. I particularly like the 6-inch Dobsonian-mounted Newtonian reflectors, like the one shown in the center photo on this page. Priced in the $400 to $650 range when properly outfitted with a finderscope and two eyepieces, these no-frills scopes are manufactured by Celestron, Meade, Orion and other companies familiar to astronomy buffs. They are available from local telescope dealers, listed in the Yellow Pages under "Telescopes,"

IN ESSENCE,
binoculars are two miniature telescopes that are capable of revealing hundreds of celestial objects invisible to the unaided eye. The powdery glow of the Milky Way is turned into a glittering river of stars in binoculars. Clumps of stars—star clusters— that are smudges to the unaided eye are transformed into stellar jewel boxes. The colors of stars are greatly enhanced through binoculars, and the four largest moons of Jupiter are brought into view.

is ideal. Binoculars magnifying beyond 10 power are too difficult to hold steady and also have narrower fields of view, though they work fine when tripod-mounted.

If you want to wear your eyeglasses while using binoculars, look for "high-eyepoint" models with large fold-down rubber eye guards. When these are folded down, the full field of view can be observed with your eyeglasses on.

An important binocular accessory for astronomy is an adapter for attaching the glasses to a camera tripod. The most common adapter (less than $20) is an L-shaped bracket with a bolt at the top of the L that

your photo dealer for a clamp tripod adapter. The difference in image clarity between handheld and tripod-mounted binoculars is astonishing.

or by mail order through the ads in the astronomy magazines.

Observing Sites *and* Light Pollution

Modern urban lighting has obliterated some—but not all—of nature's starry sky for urban dwellers

LIKE DAY AND NIGHT, *the city and country views (right), taken with the same camera, film and exposure, illustrate the impact of light pollution in a modern urban setting. Light from streetlights, office buildings, advertising signs, and so on, illuminates air molecules, dust and moisture to produce great glowing domes over our cities. Below: A non-light-polluted country site.*

W hen our grandparents were children, the splendor of a dark night sky plastered with stars and wrapped with the silky ribbon of the Milky Way was as close as the back door. From just about any backyard anywhere in the city or country, the majesty of the starry night sky was visible.

Not anymore. A giant dome of light looms over every city in North America every night. Only the brightest stars and planets punch through the glow. Anyone who lives in or near a large city sees just a few dozen stars instead of thousands. The lights of modern civilization have beaten back the stars.

Today's young people are the first generation to be born into a world where the stars are the last thing to be noticed at night instead of the first. It's a fact of life: Most of us see the stars properly only during vacation trips far from cities.

Now, I'm not suggesting that we don't need lighting at night. But a large proportion of the outdoor lighting we live with—streetlights, parking-lot illumination and so-called security lighting—is astonishingly wasteful. You can see light pollution everywhere: empty shopping-center parking lots that are floodlit all night; security lights that shine annoyingly into second-story bedroom windows; and streetlights that glare into the eyes of distant drivers instead of concentrating their light on the roadway, where it is most needed.

Many modern streetlights are designed like the fixture on the facing page, with a hemispheric glass "refractor" surrounding the light source. These ubiquitous units, little changed from when the design was introduced in the 1960s, pump up to 30 percent of their light sideways, where it illuminates nothing but the air above our heads. More efficient streetlight designs, called full-cutoff fixtures, are now available, but few municipalities use them.

With urban areas immersed in a nightly artificial twilight, one might think that interest in astronomy would be on the wane. But the reverse is emphatically the case. Sales of general-level astronomy books, astronomy magazines, telescopes and other stargazing gear are 10 to 50 times what they were in 1960, before the onslaught of urban night lighting.

The reality is that stargazing and astronomy are more popular than ever. Part of the reason is the interest stirred by a quarter-century of planetary discoveries by robot spacecraft, the recent

dramatic findings of the Hubble Space Telescope and the continuing cosmic revelations of ground-based observatories. But there is another component: the realization by more and more people that the night sky is a display of nature on the largest possible scale—a nightly pageant waiting to be observed, explored and appreciated.

RATING YOUR OBSERVING SITE

Many of us are faced with the dilemma of living where night lighting obliterates most

of the celestial wonders we want to admire. What to do? Much of the material in this book—particularly the photographic charts and accompanying descriptions—assumes that you will flee the city to seek the stars in the tranquillity of a rural or wilderness vacation. Today, this retreat from urbanization is essential to see the stars in their full glory.

But escaping from pervasive night lighting is not always possible, and you are then faced with making the best of mediocre conditions.

Depending on where you live, there might be reasonably good stargazing sites not far from your home— perhaps

even as close as your own balcony or backyard.

Observing bright objects such as the Moon and planets is possible from almost anywhere. Many strikingly beautiful conjunctions of planets and the crescent Moon (apparent close approaches to each other) occur low in the west in evening twilight and in the east in morning twilight. To see them, scout out a site at a park, conservation area, lakeshore or other such location that has a clear view in one or both of these directions.

To rate your local site for general stargazing beyond observing the sky's luminaries like the Moon and planets, select a spot protected from the direct glare of nearby lights and note whether there is any sign of the Milky Way when it is fairly high in the sky. The time to do this is after midnight in May and June, from about 11 p.m. to 1 a.m. in July, and anytime in the evening in August and September. If you can trace a hint of the Milky Way, your site is above average and suited to general stargazing in this era of widespread urban lighting.

CITY STARGAZING
The brightest nighttime objects, the Moon and naked-eye planets, can be seen reasonably well from the city. Above: Early-morning view of the crescent Moon and Mercury (left of center, just above the tallest office tower).

Crescent Moon *and* Earthshine

One of the most hauntingly beautiful nighttime sights is the thin sliver of the crescent Moon hanging in a twilight sky

cent Moon decorates the predawn sky above the eastern horizon two to five days before new Moon.

Sometimes when the Moon is a thin crescent, the rest of it is dimly visible, a ghostly glow that is particularly beautiful in binoculars. This phenomenon, traditionally called "the old Moon in the new Moon's arms," is known to modern astronomers as Earthshine.

EARTHSHINE'S SPELL

To picture what happens to produce Earthshine, imagine that you are an astronaut standing on the night side of the Moon. There is no Sun in the sky, but Earth looms as a

Swinging completely around Earth roughly once a month, our nearest celestial neighbor, the Moon, passes through a sequence of phases, from new Moon to first quarter to full to last quarter and back to new Moon. Stargazers are familiar with this cycle. But when asked to choose a favorite from these four phases, most would say, "None of the above," opting instead for

THE TWO BRIGHTEST *objects in the night sky, the Moon and Venus (right), offer a resplendent display of cosmic jewelry when they are close to each other. In this evening-sky scene, Earthshine is prominent on the Moon's night side.*

EARTHSHINE *is the illumination of the night side of the Moon by the sunlit side of Earth.*

the elegant crescent Moon.

The crescent Moon is both mysterious (hiding most of its sunlit face from our view) and elusive (the thinner it is, the closer it is to the horizon). The evening crescent Moon is seen in the western sky at nightfall from two to five days after new Moon. The morning cres-

brilliant blue-and-white globe hanging in the blackness. Because of its greater size and reflective clouds, Earth illuminates the lunar landscape 50 times more brightly than the Moon lights our nights—bright enough to make the night side visible to stargazers on Earth.

Five hundred years ago,

diamond, the planet exquisitely complements the Moon's silver sheen.

WHEN TO WATCH

Earthshine is best seen in the early evening from January through June. The favorable period for the morning sky is August to January, when the crescent Moon stands highest in northern-hemisphere skies.

Binoculars are the best optical aid for Earthshine viewing, giving better contrast than a telescope. I have frequently marveled at the phenomenon's stunning beauty in binoculars. On the sunlit crescent, the Moon's craters—some big enough to swallow the Earth's largest cities—stand in bold relief. On the night side, the detail is a subtle network of dark patches, the so-called lunar seas that are so obvious when the lunar phase advances.

Leonardo da Vinci first deduced that light from Earth can illuminate the part of the Moon not in sunlight. Earthshine varies in intensity by up to 15 percent, owing to differing amounts of cloud cover on the side of our planet facing the Moon. But it is most intense when the Moon is a thin crescent, because our satellite is then being illuminated by a nearly "full" Earth.

A crescent-Moon scene

is especially impressive if a bright planet such as Venus is nearby. Shining like a flawless

THE CRESCENT MOON *looms like a giant scimitar in this telescopic view, which reveals the cratered landscape of the alien but nearby world.*

AN EARTHSHINE MOON *is stationed below three planets in this evening-sky view (center), photographed on June 15, 1991.*

PHOTOGRAPHING THE CRESCENT MOON AND EARTHSHINE

This striking portrait of the crescent Moon and Venus in conjunction in the early-morning sky was surprisingly easy to capture on film. The trick is to be outside and waiting—and, of course, to have clear weather. The picture was taken on 100-speed film with a normal 50mm lens on a tripod-mounted standard 35mm camera. I used the camera's built-in light meter, then took several exposures at varying f-stops, slightly underexposing from the meter's reading. One of these produced the result shown at left, which includes a noticeable touch of Earthshine. The two close-up images at right, on the other hand, required a 6-inch telescope, a motorized tracking mount and a good deal of experience. They were exposed to reveal the crescent (top) and Earthshine (bottom).

Moonwatching

Binoculars easily reveal the craters and ancient lava plains covering the face of our nearest celestial neighbor

THE SAME SIDE *of the Moon faces Earth at all times, no matter what the phase or time of year or where it is in its orbit. The Moon is in what astronomers call rotational lock, with its slightly more massive side firmly in the Earth's gravitational grip and permanently aligned Earthward. Thus the pictures here, showing first quarter (facing page) and last quarter (right) combine to match the image of the full Moon below.*

Mare Frigoris (Sea of Cold)

Plato

Sinus Iridum (Bay of Rainbows)

Aristillus

Mare Imbrium (Sea of Rains)

Autolycus

Timocharis

Archimedes

Aristarchus

Apennine Mts.

Oceanus Procellarum (Ocean of Storms)

Eratosthenes

Copernicus

Sinus Aestuum (Seething Bay)

Kepler

Sinus Medii (Central Bay)

Reinhold

Ptolemaeus

Grimaldi

Alphonsus

Arzachel

Straight Wall

Mare Humorum (Sea of Moisture)

Mare Nubium (Sea of Clouds)

Tycho

Longomontanus

Maginus

Clavius

The Moon was one of the first objects observed after the invention of the telescope in 1609. Astronomers of the time thought that the smooth dark regions might be the lunar equivalent of the Earth's oceans and named them accordingly. Later, the craters were named after philosophers and astronomers of the past.

The craters are the remnants of comet and asteroid collisions that occurred during the first billion years after the solar system's formation. Since then, the Moon has changed little. The lunar "seas" are the remains of large hits that unleashed lava flows which eventually solidified into the Moon's distinctive dark plains visible to the naked eye.

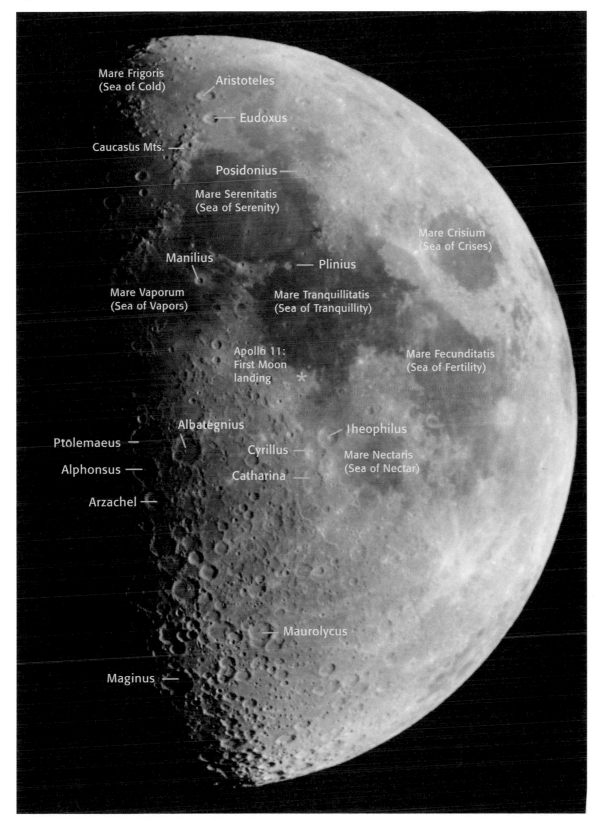

Mare Frigoris (Sea of Cold)
Aristoteles
Eudoxus
Caucasus Mts.
Posidonius
Mare Serenitatis (Sea of Serenity)
Mare Crisium (Sea of Crises)
Manilius
Plinius
Mare Vaporum (Sea of Vapors)
Mare Tranquillitatis (Sea of Tranquillity)
Apollo 11: First Moon landing
Mare Fecunditatis (Sea of Fertility)
Albategnius
Theophilus
Ptolemaeus
Cyrillus
Mare Nectaris (Sea of Nectar)
Alphonsus
Catharina
Arzachel
Maurolycus
Maginus

THE BEST TIME
for moonwatching is from four days before to three days after first quarter, when craters and mountains are thrown into stark relief along the terminator, or shadow line. Binoculars, particularly if tripod-mounted, will show all the features identified here. Large craters have the following diameters: Copernicus, 93 kilometers (58 mi); Tycho, 85 kilometers (53 mi); Clavius, 225 kilometers (140 mi); Theophilus, 100 kilometers (62 mi); Plato, 101 kilometers (63 mi); Ptolemaeus, 153 kilometers (95 mi). At 4,850 meters (16,000 ft) from floor to rim, Tycho is one of the deepest craters.

Observing *the* Planets

The brightest planets are fascinating to watch as they trek through the constellations

JUPITER'S FOUR LARGEST *moons (above) are visible in ordinary binoculars. The drawing at upper right shows the planet's cloud belts and the shadow of a moon as seen in an amateur astronomer's telescope. Photo below records the June 1991 grouping of Venus (brightest), Jupiter and Mars (faintest).*

Like the Moon, planets shine by reflected sunlight. Although they are hundreds of times farther away than the Moon, the planets are still in our own celestial neighborhood, and they often appear brighter than any stars, as illustrated in the photos on these two pages. Venus, the planet that shines brightest in our sky, is dramatically more luminous than any star and is therefore easily visible from the city in twilight. It's the object below and to the left of the crescent Moon in the center photo on this page.

But brightness alone is not enough to determine whether you are looking at a planet

or a bright star. The following guidelines will help you distinguish one from the other.

First, unlike stars, planets seldom twinkle. The reason

for this difference in their appearance tells a lot about the difference between planets and stars themselves.

Stars twinkle because they are pinpoint light sources. Of course, the stars are far from tiny in reality, but their enormous distances reduce them to disks so minuscule that even the largest telescopes cannot reveal them. They are, in effect, point sources of light. Thus a beam of starlight entering your eye is a fragile cosmic thread that is easily rippled by the ever present tur-

bulence in the Earth's atmosphere, which causes the star to twinkle.

A planet, however, is not a pinpoint; it's close enough to

show a perceptible disk. Planetary disks are a bit too small for the unaided eye to see, but ordinary binoculars will reveal them as noticeably fatter than stars. The bigger bundle of light from the planetary disk is less easily disrupted

by atmospheric turbulence and therefore looks steady and unwavering (though exceptionally turbulent air, especially when a planet is close to the horizon, can cause vigorous twinkling).

ZODIAC BELT

Although planets move, they are always confined to the ecliptic, the plane of the solar system, as explained in the box on facing page. The ecliptic is the second guideline used to identify planets.

The ecliptic touches the 12 constellations of the zodiac, a belt that includes such well-known stellar groups as Leo, Scorpius, Virgo and Libra. Thousands of years ago, skywatchers elevated the zodiac constellations to special status. Then, they were regarded as a royal pathway for the magical

THE ECLIPTIC: PLANETARY PATHWAY

The Sun, Moon and planets never stray from a well-defined heavenly boulevard called the ecliptic. You can think of the ecliptic as the top of a billiard table, with the planets represented by balls moving across its surface. Of course, the Earth is one of those balls too, so our viewpoint is from the table looking around at the other balls. This arrangement exists because the solar system is essentially flat. The planets all orbit in the same direction and in the same plane (see illustration on page 7). In the photo at right, taken in May 1991, three planets and the Moon clearly define the ecliptic's orientation in the sky. In effect, the ecliptic is the solar system's equator. The ecliptic's angle relative to the horizon is an effect of latitude; the farther you are from the equator, the lower the ecliptic is toward the horizon. Also, the angle varies with the time of year. In the west after sunset—the direction most favored for planet spotting—the ecliptic is angled highest from January to June and is less well positioned for the remainder of the year. The seasonal shift is evident in the all-sky charts on pages 20 and 22 and 38-39, where the ecliptic's position is marked. Why are the planetary orbits in a flat plane instead of randomly angled? The pancake orientation seems to date from the solar system's formation nearly five billion years ago, when the Sun and its family of planets were born from a flat cloud of dust and gas. Since the primordial material was a thin disk, the orbital arrangement of the planets has retained that basic shape to this day.

PLANETARY ROLL CALL

Each of the five planets that are visible to the unaided eye has its own individual characteristics. These are outlined on

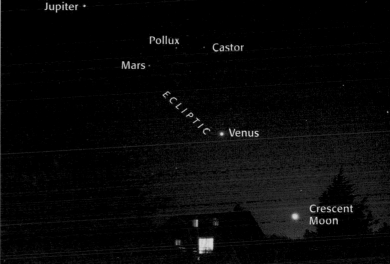

page 60, where you will find tables of their locations for the current year.

The four brightest and most easily observed planets—Venus, Jupiter, Saturn and Mars—are endlessly fascinating for the recreational astronomer. The orbital clockwork that governs their motions never repeats exactly, so their wanderings are always fresh and fun to follow from week to week.

You will especially want to monitor the evening and morning twilight skies, where the maximum planet action takes place, including truly inspiring conjunctions of planets and the crescent Moon, as shown here and on pages 12 and 13.

Notice the difference in the way the ecliptic is angled in the morning and evening skies, as illustrated in the two photos on this page. Evening facing west (above), the ecliptic angles up and to the left. Morning facing east (left), the ecliptic angles up and to the right. In both cases, the Moon and planets line up along this well-worn celestial pathway.

PLANETS ORBIT
the Sun just as Earth does and shine by reflected sunlight, as does the Moon. Because they are in our celestial neighborhood, planets are often brighter than any stars, as can be seen in the photo above. Over a period of days or weeks, the orbital motion of the planets (including Earth) shifts them relative to the background stars.

"moving stars"—the planets. Our forebears had no idea why the planets moved or why they trekked only through certain star groups, and they spent a lot of time speculating about the phenomenon. From this well of ignorance emerged horoscopes and predictions about the lives and futures of royalty as well as peasants—a practice known as astrology that continues to this day (with special emphasis on movie and television stars). The science of astronomy long ago shed any association with such fortune-telling.

How *to* Find Stars *and* Constellations

The Big Dipper is the stargazer's premier guidepost, pointing the way to key stars and constellations

THE EARTH'S ROTATION *on its axis causes the stars to trail in this one-hour time-exposure photo. Polaris, the North Star, is the bright object close to center.*

Although the sky on a dark night seems thronged with stars, the number of stars seen as individuals by the unaided eye is fewer than 4,000. It just looks like millions. But even 4,000 is a daunting number when you are trying to get to know them.

The key to star and constellation identification is to start with the night sky's most prominent guidepost, the Big Dipper. You probably know the Big Dipper already, but if you are unsure, look at the diagram below, which relates the Dipper's size to your hand held at arm's length.

From May to October, the Big Dipper is seen in the north-northwest throughout North America. The three-panel illustration on facing page shows its mid-evening location relative to the horizon and overhead throughout this period. These are very wide-angle views encompassing about one-third of the sky. It's pretty obvious that the Big Dipper is a very efficient guidepost—an attribute which will become even more apparent as you proceed through the chart pages that follow.

Initially, though, I recommend spending one or two evenings outside, establishing

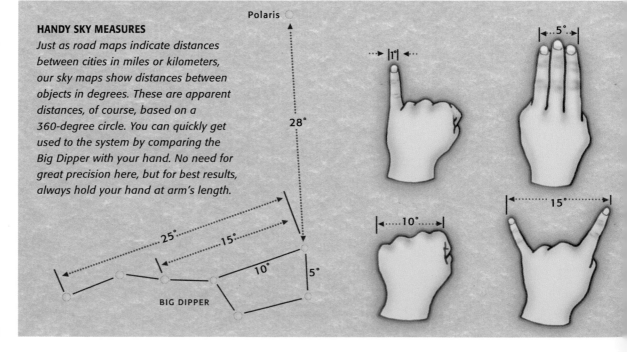

HANDY SKY MEASURES
Just as road maps indicate distances between cities in miles or kilometers, our sky maps show distances between objects in degrees. These are apparent distances, of course, based on a 360-degree circle. You can quickly get used to the system by comparing the Big Dipper with your hand. No need for great precision here, but for best results, always hold your hand at arm's length.

Polaris

28°

25°   15°   10°   5°

1°   5°   10°   15°

BIG DIPPER

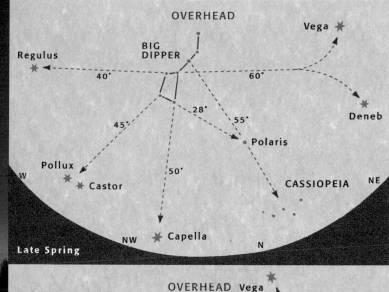

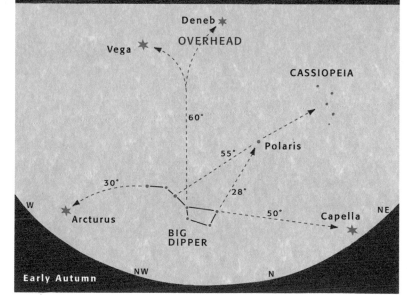

the Big Dipper's key pointing system and committing as much of it to memory as possible. You will be using this locator-arrow system every time you stargaze. It is the fundamental framework upon which your stargazing skills will be built. And since it involves only the sky's brightest stars, your training sessions can be done anywhere but the most light-polluted city locations.

(On the mid-summer chart, the Dipper's handle points to the bright stars Arcturus and Spica. This pointer also applies to the late-spring chart, but these stars are "past" overhead and out of the frame, so they are not shown.)

SHIFTING SKIES

The three charts at right show a gradual counterclockwise shift of the whole sky as the seasons progress. This is caused by the Earth's passage along its orbit around the Sun. The cycle repeats itself every year.

Another sky shift occurs far more rapidly. As Earth turns on its axis, the sky "rotates" in the opposite direction. As a result, we see the Sun rise in the east and set in the west. Stars also rise in the east and

set in the west. In the northern sector of the sky, though, they swing counterclockwise around Polaris, the North Star. These circumpolar stars are in view whenever the sky is dark. Fortunately for the United States and Canada, the Big Dipper stars are also circumpolar and are therefore always available for reference.

WHAT ARE STARS?

Stars are suns basically like our own Sun. As the tiny spears of light from those remote suns enter the eyes of a stargazer on Earth, a journey that began long ago comes to an end. That is ancient light we see. For example, five of the seven stars in the Big Dipper are about 75 light-years away. The light from these stars takes a human lifetime to reach Earth.

Although the stars are moving relative to one another, the vast distances between them mean that the change in position is so ponderous as to be

completely undetectable over several human life spans. We see exactly the same stars in the same place as those seen by our grandparents and their grandparents before them.

MOONLIT SCENE
at top shows the Big Dipper and Polaris.

19

Key Chart: Late Spring *and* Early Summer

The major stars and constellations visible from the United States and Canada from late April to the end of June

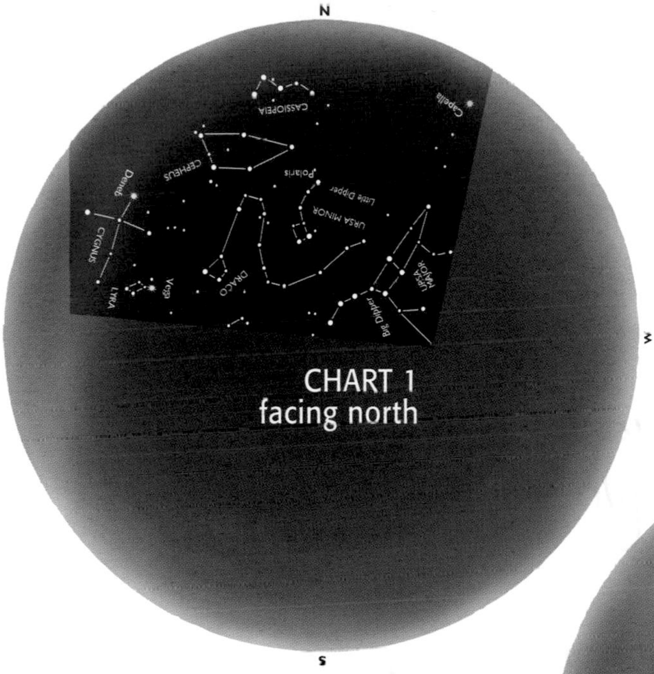

CHART 1
facing north

N

CASSIOPEIA

CEPHEUS

Polaris

Deneb

CYGNUS

URSA MINOR

LYRA

VEGA

DRACO

Big Dipper

URSA MAJOR

Capella

W

S

These four sectors of the Key Chart on facing page are the areas shown in detail on Charts 1, 2, 3 and 4 (pages 24-37).

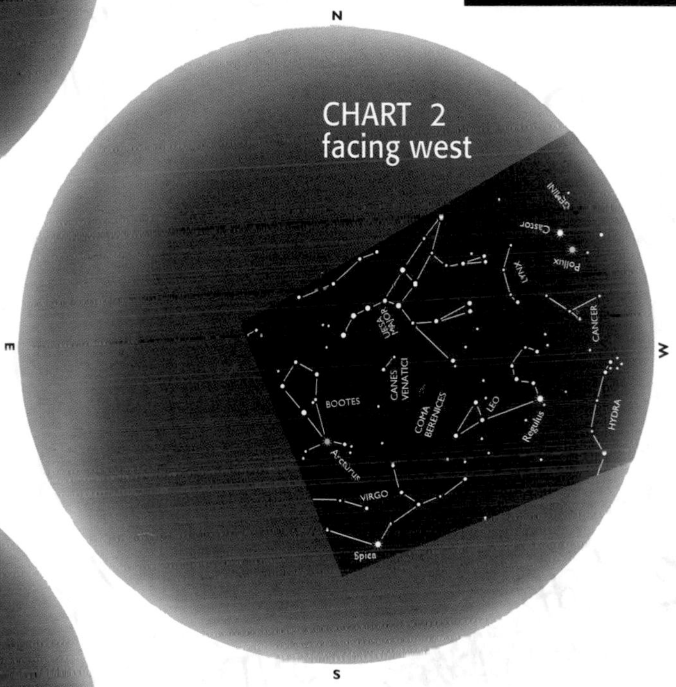

CHART 2
facing west

N

URSA MAJOR

CANES VENATICI

COMA BERENICES

LEO

Regulus

CANCER

Pollux

Castor

LYNX

Capella

BOOTES

Arcturus

VIRGO

Spica

HYDRA

m

W

S

CHART 4
facing east

N

Deneb

CYGNUS

LYRA

Vega

DRACO

URSA MAJOR

Big Dipper

Altair

AQUILA

HERCULES

CORONA BOREALIS

BOOTES

CANES VENATICI

SERPENS CAPUT

OPHIUCHUS

Arcturus

VIRGO

W

S

CHART 3
facing south

N

URSA MAJOR

BOOTES

CANES VENATICI

COMA BERENICES

LEO

CORONA BOREALIS

SERPENS CAPUT

Arcturus

VIRGO

Spica

CORVUS

LIBRA

SCORPIUS

Antares

CENTAURUS

HYDRA

m

W

S

CIRCULAR STAR CHART on facing page and the charts on pages 22, 48 and 49 show the entire sky visible at the times indicated. The chart's edge is the horizon; the overhead point is at center. The chart is most effective when you use about one-quarter of it at a time—a roughly pie-shaped wedge—which approximates a comfortable naked-eye field of view in a given direction. Outdoors, match the horizon compass direction on the chart with the actual direction you are facing. Don't be confused by the east and west points on the chart being opposite to their locations on a map of Earth. When the chart is held up to match the sky, with the direction you are facing at the bottom, the chart directions will match the compass points. (Continued on page 23.)

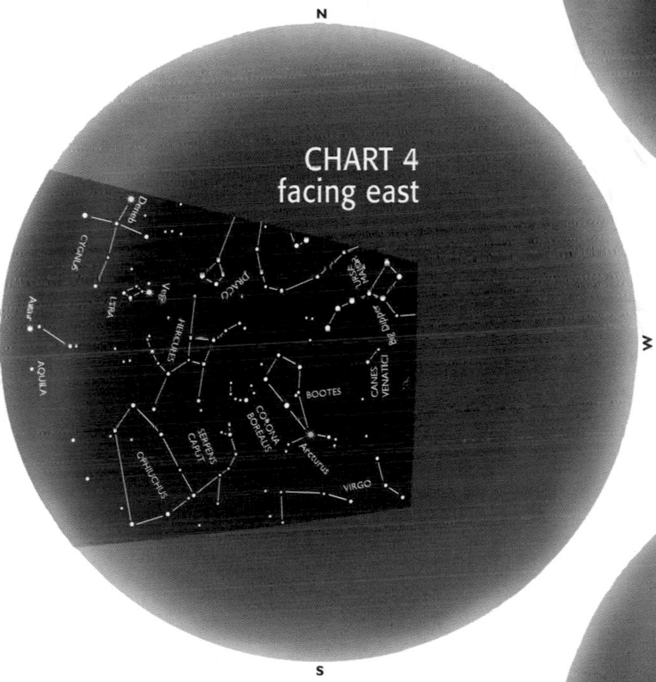

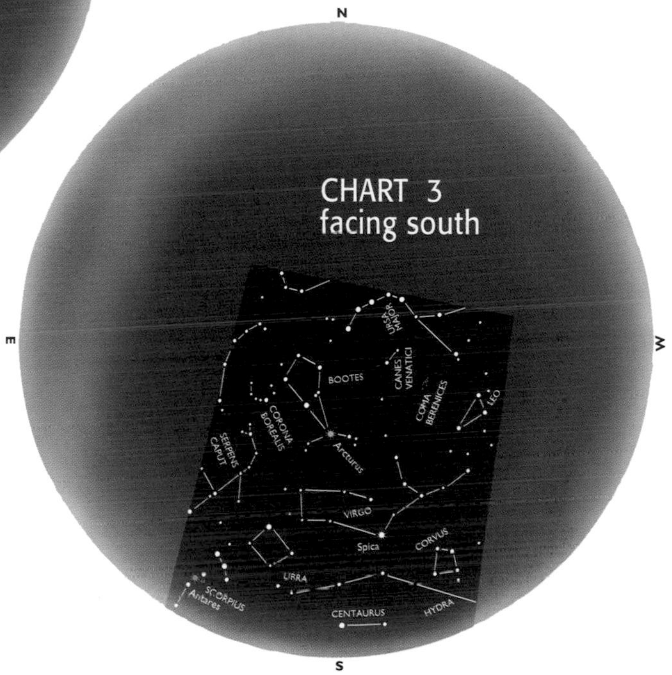

Use Key Chart at left within one hour of these times (local daylight time).

Late April	**midnight**
Early May	**11 pm**
Late May	**10 pm**
Early June	**9 pm**
Late June	**dusk**

The major stars and constellations visible from the United States and Canada from July to September

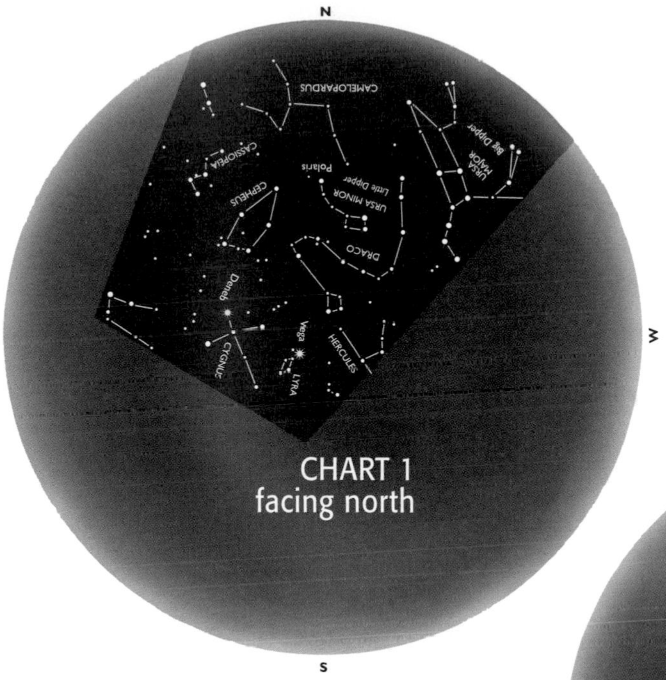

CHART 1
facing north

These four sectors of the Key Chart on facing page are the areas shown in detail on Charts 1, 5, 6 and 7 (pages 24-37).

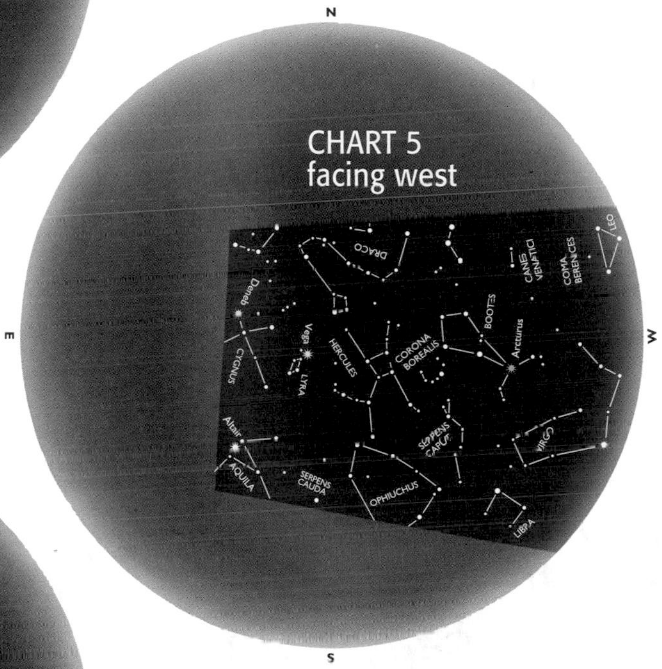

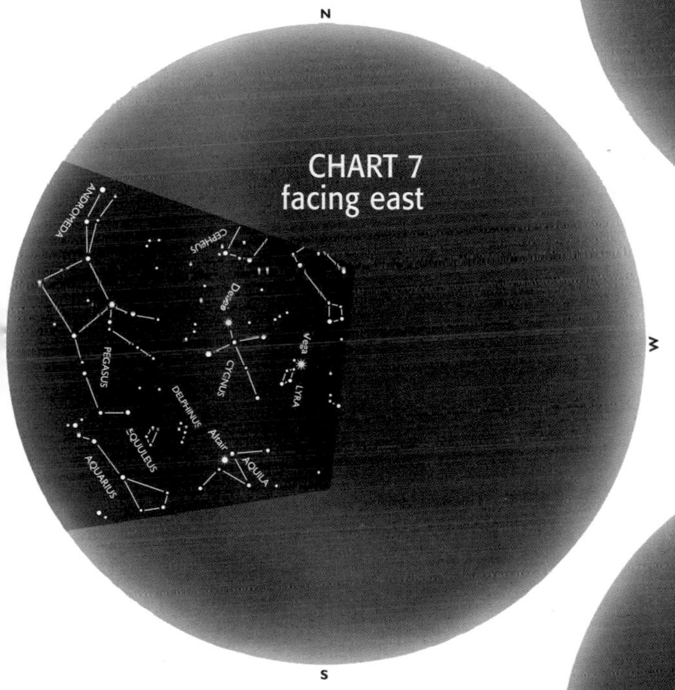

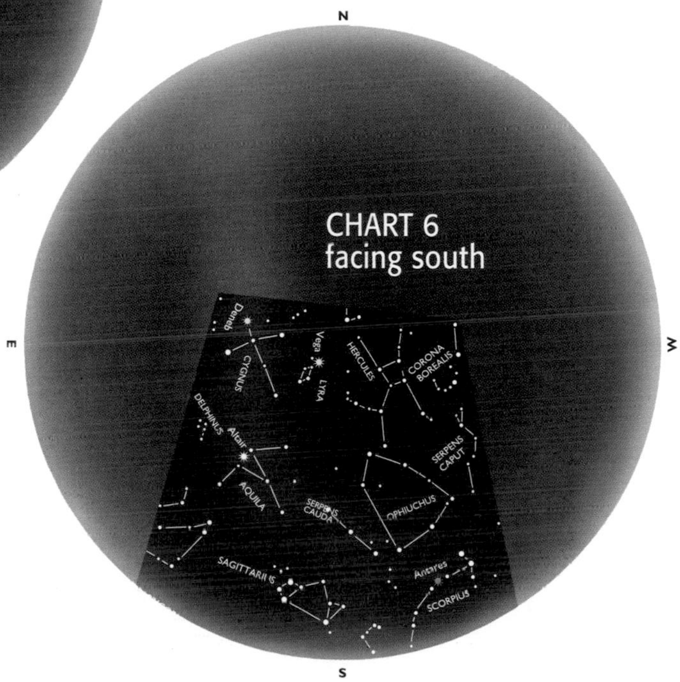

CHART 5
facing west

CHART 7
facing east

CHART 6
facing south

CIRCULAR STAR CHART on facing page and the charts on pages 20, 48 and 49 show star groups joined by lines that outline the constellations created by our ancestors thousands of years ago. These figures were used to map the night sky. Modern astronomers continue to use the traditional names (always capitalized on our charts), providing today's stargazers with a permanent link to the sky myths and legends of the past. The ecliptic is the celestial pathway of the Moon and planets (see page 17). The star groups straddling this line are known as the zodiac constellations. When referring to the charts outdoors, use a flashlight heavily dimmed with red plastic or cellophane. Unfiltered lights significantly reduce your night-vision sensitivity. On a moonless night in the country, you will see more stars than are shown here; within a city, you will see fewer. For instructions on the use of circular star charts, see page 21.

Use Key Chart at left within one hour of these times (local daylight time).

Early July	midnight
Late July	11 pm
Early August	10 pm
Late August	9 pm
Early September	dusk
Late September	dusk

Known as circumpolar stars, these constellations are visible throughout the year from most of North America.

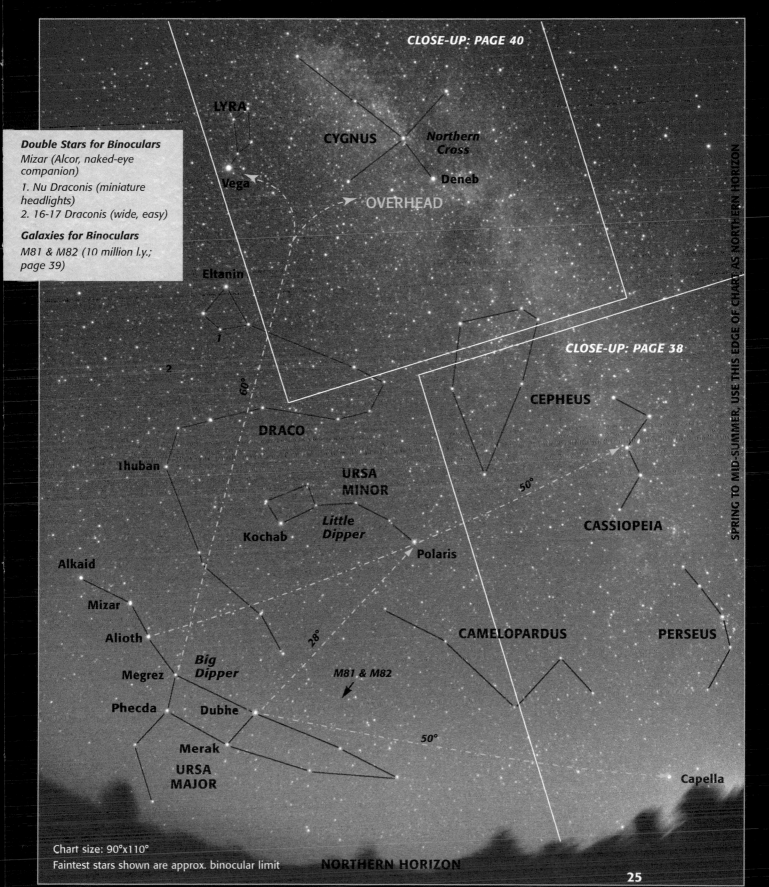

CLOSE-UP: PAGE 40

LYRA

CYGNUS

Northern Cross

Deneb

OVERHEAD

Double Stars for Binoculars
Mizar (Alcor, naked-eye companion)
1. Nu Draconis (miniature headlights)
2. 16-17 Draconis (wide, easy)

Galaxies for Binoculars
M81 & M82 (10 million l.y.; page 39)

Vega

Eltanin

1

2

60°

DRACO

CLOSE-UP: PAGE 38

CEPHEUS

Thuban

URSA MINOR

50°

CASSIOPEIA

Little Dipper

Kochab

Polaris

Alkaid

Mizar

Alioth

28°

CAMELOPARDUS

PERSEUS

Big Dipper

Megrez

M81 & M82

Phecda

Dubhe

50°

Merak

URSA MAJOR

Capella

Chart size: 90°x110°
Faintest stars shown are approx. binocular limit

NORTHERN HORIZON

SPRING TO MID-SUMMER, USE THIS EDGE OF CHART AS NORTHERN HORIZON

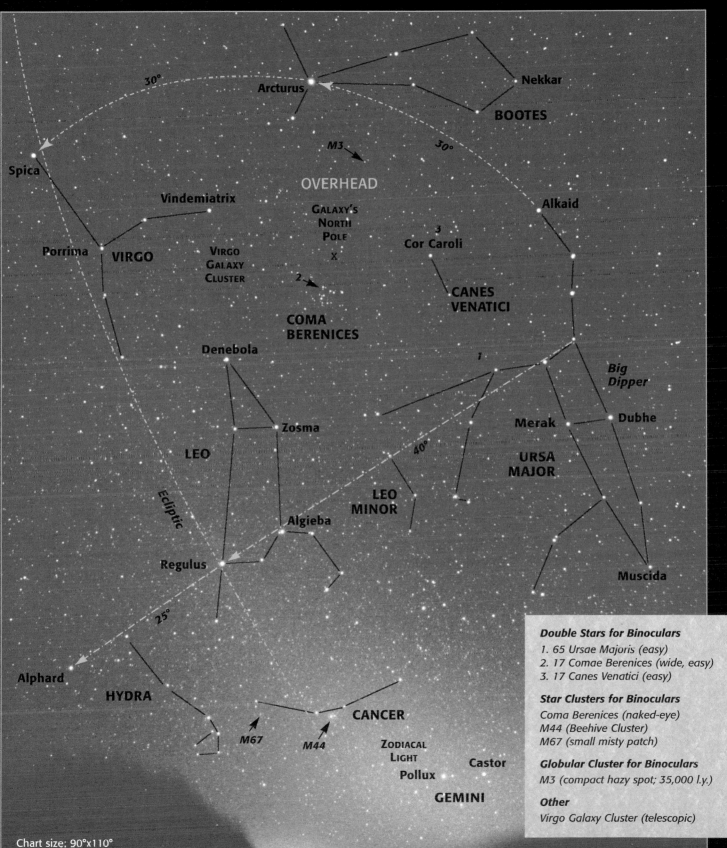

30°

Arcturus

Nekkar

BOOTES

30°

M3

OVERHEAD

Spica

Alkaid

Vindemiatrix

GALAXY'S
NORTH
POLE

3

X

Cor Caroli

Porrima

VIRGO

VIRGO
GALAXY
CLUSTER

2

CANES
VENATICI

COMA
BERENICES

1

Big
Dipper

Denebola

Zosma

Merak

Dubhe

LEO

URSA
MAJOR

Ecliptic

LEO
MINOR

40°

Algieba

Regulus

Muscida

25°

Alphard

HYDRA

M67

M44

CANCER

ZODIACAL
LIGHT

Castor

Pollux

GEMINI

Double Stars for Binoculars

1. 65 Ursae Majoris (easy)
2. 17 Comae Berenices (wide, easy)
3. 17 Canes Venatici (easy)

Star Clusters for Binoculars

Coma Berenices (naked-eye)
M44 (Beehive Cluster)
M67 (small misty patch)

Globular Cluster for Binoculars

M3 (compact hazy spot; 35,000 l.y.)

Other

Virgo Galaxy Cluster (telescopic)

Chart size: 90°x110°
Faintest stars shown are approx. binocular limit

WESTERN HORIZON

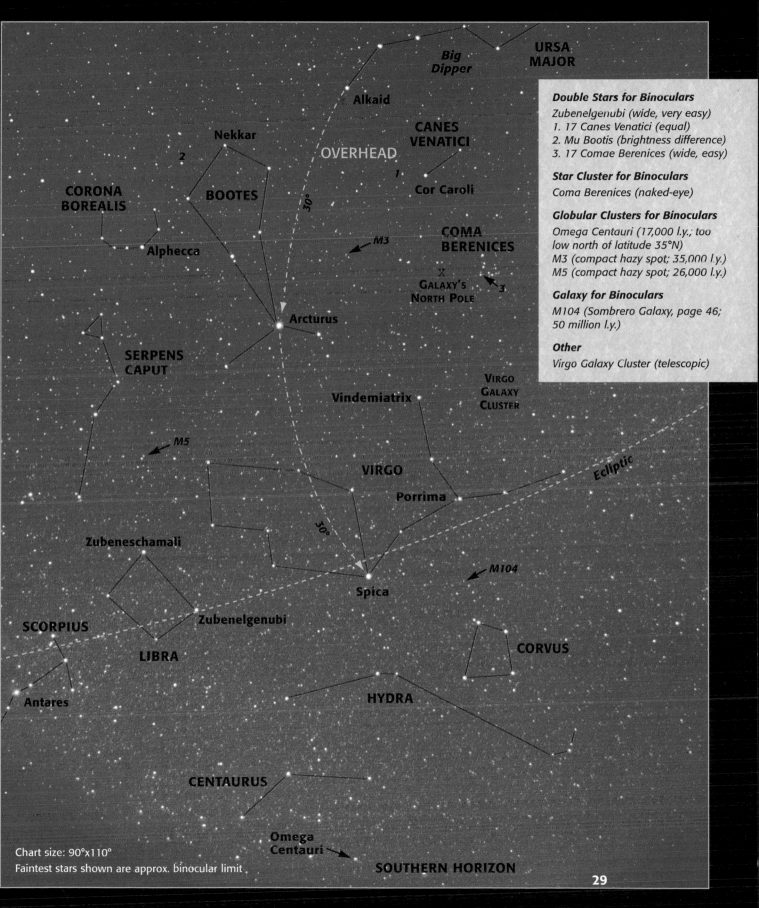

Double Stars for Binoculars
Zubenelgenubi (wide, very easy)
1. 17 Canes Venatici (equal)
2. Mu Bootis (brightness difference)
3. 17 Comae Berenices (wide, easy)

Star Cluster for Binoculars
Coma Berenices (naked-eye)

Globular Clusters for Binoculars
Omega Centauri (17,000 l.y.; too low north of latitude 35°N)
M3 (compact hazy spot; 35,000 l.y.)
M5 (compact hazy spot; 26,000 l.y.)

Galaxy for Binoculars
M104 (Sombrero Galaxy, page 46; 50 million l.y.)

Other
Virgo Galaxy Cluster (telescopic)

Chart size: 90°x110°
Faintest stars shown are approx. binocular limit

Hercules and Corona Borealis, the Northern Crown, are easily found along a line from Arcturus to Vega.

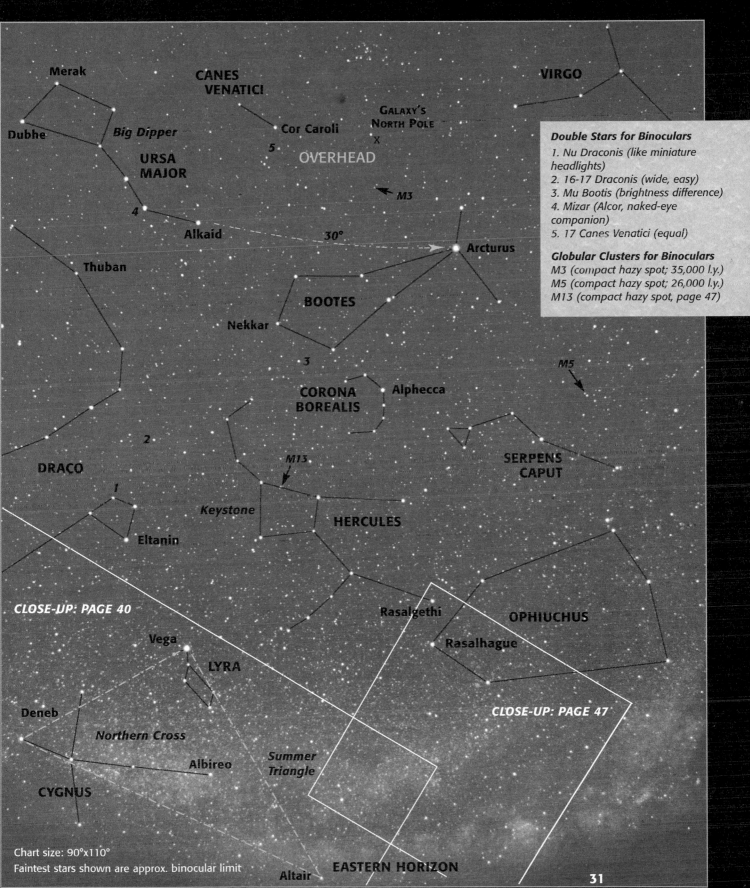

Merak

CANES VENATICI

VIRGO

Dubhe

Cor Caroli

GALAXY'S NORTH POLE
X

Big Dipper

5

OVERHEAD

URSA MAJOR

M3

4

Alkaid

30°

Arcturus

Thuban

BOOTES

Nekkar

3

CORONA BOREALIS

Alphecca

M5

2

DRACO

SERPENS CAPUT

M13

1

Keystone

Eltanin

HERCULES

CLOSE-UP: PAGE 40

Rasalgethi

OPHIUCHUS

Vega

Rasalhague

LYRA

Deneb

CLOSE-UP: PAGE 47

Northern Cross

Albireo

Summer Triangle

CYGNUS

Chart size: 90°x110°
Faintest stars shown are approx. binocular limit

Altair

EASTERN HORIZON

> **Double Stars for Binoculars**
> *1. Nu Draconis (like miniature headlights)*
> *2. 16-17 Draconis (wide, easy)*
> *3. Mu Bootis (brightness difference)*
> *4. Mizar (Alcor, naked-eye companion)*
> *5. 17 Canes Venatici (equal)*
>
> **Globular Clusters for Binoculars**
> M3 (compact hazy spot; 35,000 l.y.)
> M5 (compact hazy spot; 26,000 l.y.)
> M13 (compact hazy spot, page 47)

As the Milky Way approaches overhead in mid- to late summer,
Arcturus and the stars of spring slowly sink in the west.

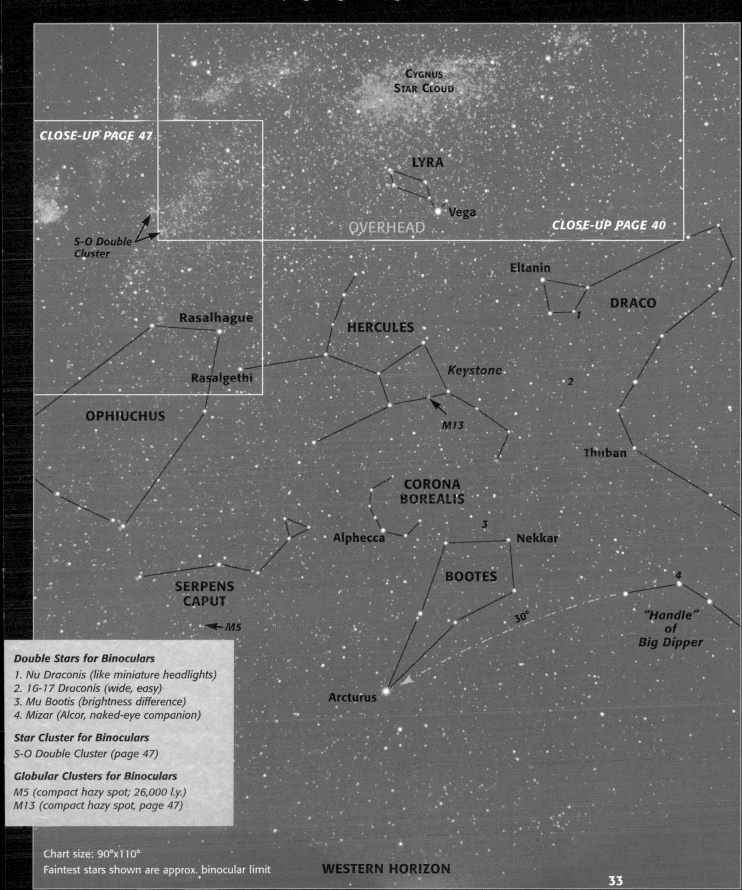

CYGNUS
STAR CLOUD

CLOSE-UP PAGE 47

LYRA

Vega

OVERHEAD

CLOSE-UP PAGE 40

S-O Double
Cluster

Eltanin

DRACO

1

Rasalhague

HERCULES

Keystone

Rasalgethi

2

M13

OPHIUCHUS

Thuban

CORONA
BOREALIS

3

Nekkar

Alphecca

BOOTES

4

SERPENS
CAPUT

30°

"Handle"
of
Big Dipper

M5

Arcturus

Double Stars for Binoculars
1. Nu Draconis (like miniature headlights)
2. 16-17 Draconis (wide, easy)
3. Mu Bootis (brightness difference)
4. Mizar (Alcor, naked-eye companion)

Star Cluster for Binoculars
S-O Double Cluster (page 47)

Globular Clusters for Binoculars
M5 (compact hazy spot; 26,000 l.y.)
M13 (compact hazy spot, page 47)

Chart size: 90°x110°
Faintest stars shown are approx. binocular limit

WESTERN HORIZON

The summer Milky Way blazes forth from mid-July through September as we peer toward the galaxy's core.

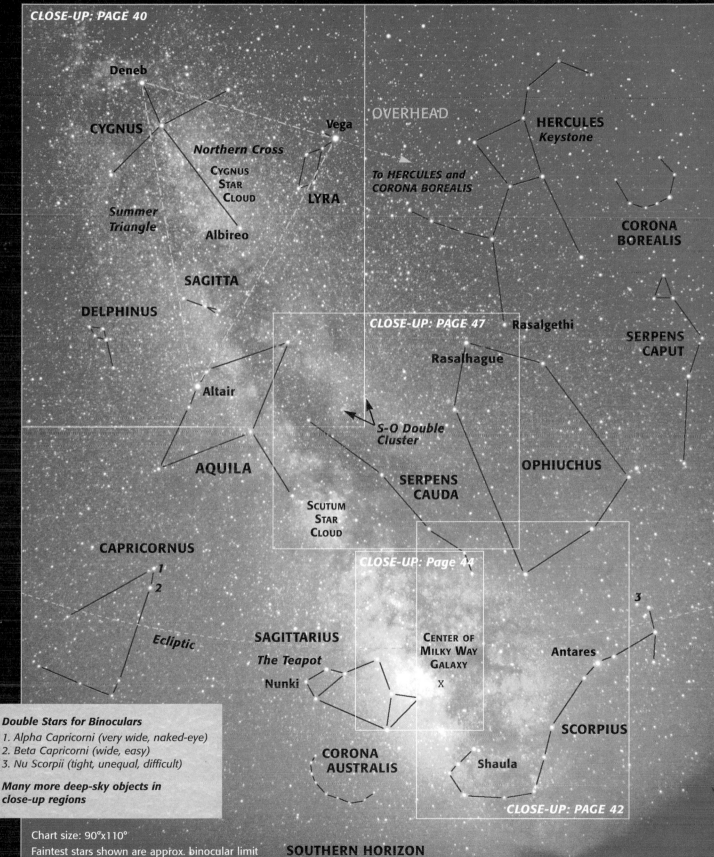

CLOSE-UP: PAGE 40

Deneb

CYGNUS

OVERHEAD

HERCULES
Keystone

Vega

Northern Cross

CYGNUS STAR CLOUD

LYRA

To HERCULES and CORONA BOREALIS

Summer Triangle

Albireo

CORONA BOREALIS

SAGITTA

DELPHINUS

CLOSE-UP: PAGE 47   Rasalgethi

SERPENS CAPUT

Rasalhague

Altair

S-O Double Cluster

OPHIUCHUS

AQUILA

SERPENS CAUDA

SCUTUM STAR CLOUD

CAPRICORNUS

1
2

CLOSE-UP: Page 44

3

Ecliptic

SAGITTARIUS

The Teapot

Nunki

CENTER OF MILKY WAY GALAXY

X

Antares

SCORPIUS

Double Stars for Binoculars
1. Alpha Capricorni (very wide, naked-eye)
2. Beta Capricorni (wide, easy)
3. Nu Scorpii (tight, unequal, difficult)

Many more deep-sky objects in close-up regions

CORONA AUSTRALIS

Shaula

CLOSE-UP: PAGE 42

Chart size: 90°x110°
Faintest stars shown are approx. binocular limit

SOUTHERN HORIZON

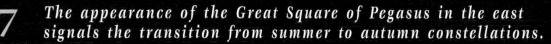

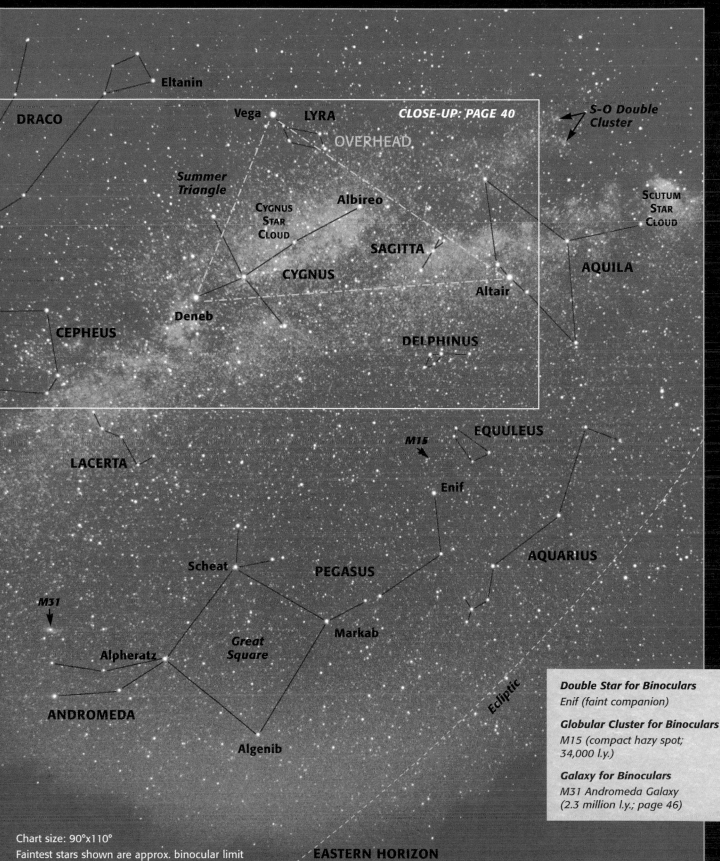

Eltanin

DRACO

Vega LYRA OVERHEAD *CLOSE-UP: PAGE 40*

S-O Double Cluster

Summer Triangle

CYGNUS STAR CLOUD Albireo

SCUTUM STAR CLOUD

SAGITTA

CYGNUS

AQUILA

Altair

Deneb

CEPHEUS

DELPHINUS

EQUULEUS

M15

LACERTA

Enif

AQUARIUS

Scheat PEGASUS

M31

Markab

Alpheratz *Great Square*

Ecliptic

ANDROMEDA

Algenib

Double Star for Binoculars
Enif (faint companion)

Globular Cluster for Binoculars
M15 (compact hazy spot; 34,000 l.y.)

Galaxy for Binoculars
M31 Andromeda Galaxy (2.3 million l.y.; page 46)

Chart size: 90°x110°
Faintest stars shown are approx. binocular limit

EASTERN HORIZON

CASSIOPEIA
is second only to the Big Dipper as a guide-post in the northern sector of the night sky. You'll see it in the northeast, as it appears in this photo, between midnight and 2 a.m. in July; between 10 p.m. and midnight during August; and in the early evening in September and October. The constellation's distinctive 15-degree-wide W pattern, delineated by stars almost as bright as those of the Big Dipper, makes finding it a snap. Two of the best deep-sky treasures are nearby. Use binoculars to locate the Double Cluster, about midway between the central star of the W and Mirfak, the brightest star in Perseus. The Andromeda Galaxy is 15 degrees to the right of the W, on the opposite side from Polaris. The Perseus Cluster, surrounding Mirfak, is a sparkling splash of stars in binoculars.

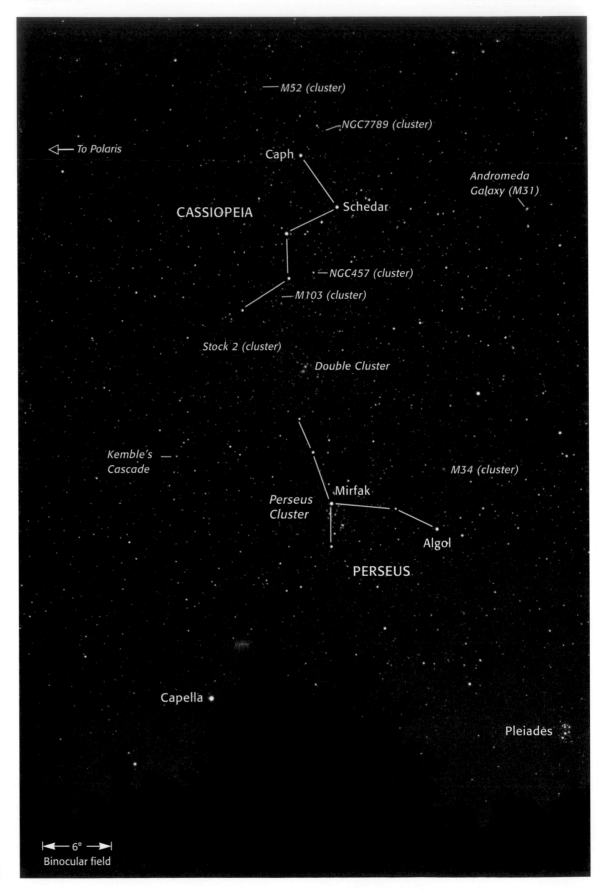

The section of the Milky Way Galaxy running through Cassiopeia and Perseus is the next spiral arm outward from the arm of the galaxy in which our Sun resides. A string of star clusters decorates this galactic territory, led by the **Double Cluster** in Perseus, visible as a hazy patch to the naked eye. It is one of the great

Double Cluster

sights in binoculars and back-yard telescopes. Binoculars re-veal dozens of individual stars in each of these congregations, even though they are 7,000 light-years away. In the same binocular field as the Double Cluster is **Stock 2**, a large dispersed cluster about 1,000 light-years distant.

The other clusters in and

M81

M82

around Cassiopeia pale by com-parison to the Double Cluster, but each is worth tracking down. The best of this second-tier group is **NGC457**.

A locator arrow extended 15 degrees off the "flat" end of the W of Cassiopeia will direct you to **Kemble's Cascade**, a neat chain of stars visible in binoc-

ulars that ends at a small clus-ter. Such a pleasing chance grouping of stars that are other-wise unrelated is known as an asterism.

The area around the Big Dipper contains four of the sky's largest and brightest galaxies, though binocular and small-telescope users will find them anything but big and bright. Galaxies the size of the Milky Way are reduced to puny smudges when seen from distances of millions of light-years. Be patient, wait for a clear, moonless night, and then prepare by let-ting your eyes become

M51

dark-adapted for at least 10 minutes. **M81** should be your first target. It is the third brightest galaxy visible from

North America, after the Andromeda Galaxy (M31) and M33. In many ways, however, it is the most impressive, because less than a degree away is **M82**, and the shapes of both galaxies are evident in binoculars. This galaxy duo is 11 million light-years away.

M101, at a distance of 20 million light-years, appears larger yet dimmer than M81. **M51**, more distant still, at 25 million light-years, is a small, hazy patch that is chal-lenging to identify in binocu-lars. Difficult they may be, but what a thrill to see so far with such humble equipment!

Kemble's Cascade

THE BIG DIPPER
(below) and its much less prominent junior partner, the Little Dipper, are both defined by seven stars. Four galaxies visible in binoculars surround the Big Dipper.

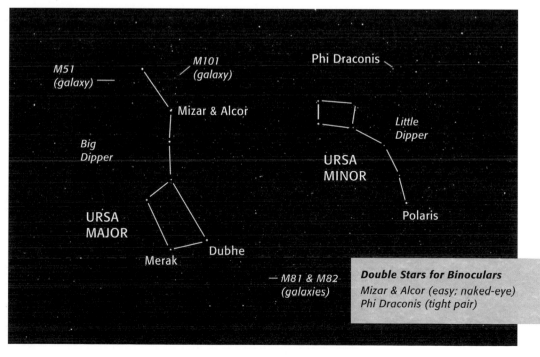

M51
(galaxy)

M101
(galaxy)

Phi Draconis

Mizar & Alcor

Big Dipper

Little Dipper

URSA MINOR

URSA MAJOR

Merak

Dubhe

Polaris

— M81 & M82
(galaxies)

Double Stars for Binoculars
Mizar & Alcor (easy; naked-eye)
Phi Draconis (tight pair)

THE NORTHERN CROSS, *traditionally known as Cygnus the Swan, straddles the summer Milky Way and includes some of the best stargazing territory the sky has to offer. The central plane of the Milky Way Galaxy runs very close to a line from Deneb to Albireo. An especially rich portion of the Milky Way known as the Cygnus Star Cloud is brimming with stars just at the threshold of naked-eye visibility, which become transformed into a stellar wonderland in binoculars. The Cygnus Star Chain is a conspicuous clumping of the myriad stars in this area. Scan with binoculars along a line from Altair to Vega, and you will encounter the Coathanger Cluster, a delightful stellar grouping that looks just like its name.*

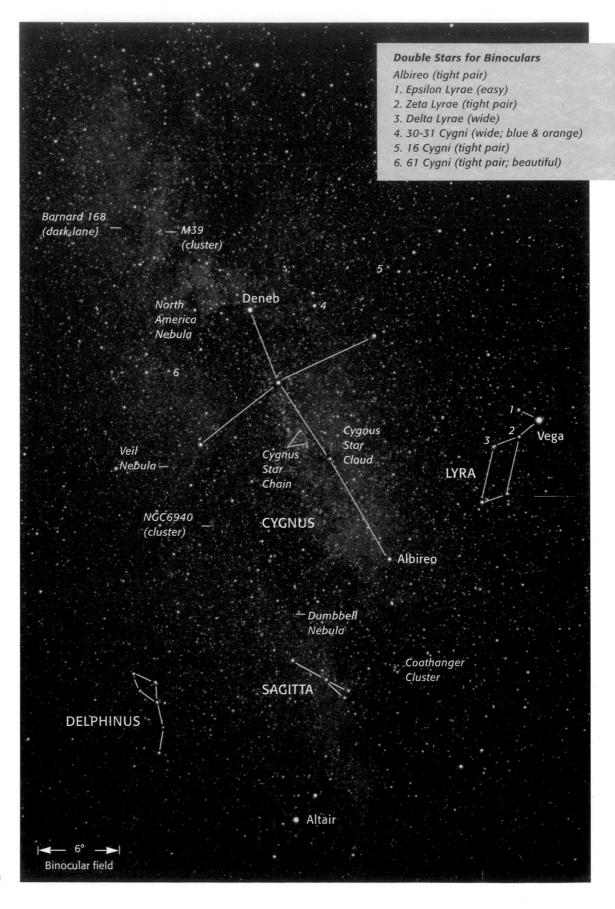

Double Stars for Binoculars
Albireo (tight pair)
1. Epsilon Lyrae (easy)
2. Zeta Lyrae (tight pair)
3. Delta Lyrae (wide)
4. 30-31 Cygni (wide; blue & orange)
5. 16 Cygni (tight pair)
6. 61 Cygni (tight pair; beautiful)

Barnard 168 (dark lane)
M39 (cluster)
North America Nebula
Deneb
5
4
6
Cygnus Star Cloud
Cygnus Star Chain
Veil Nebula
1
2
3
Vega
LYRA
NGC6940 (cluster)
CYGNUS
Albireo
Dumbbell Nebula
Coathanger Cluster
SAGITTA
DELPHINUS
Altair

6°
Binocular field

Dumbbell Nebula

North America Nebula and Deneb

The irregular, patchy appearance of the Milky Way stems from its dual composition: stars, and dusty gas clouds yet to form into stars. The stars shine, while the dust and gas conceal. This combination produces features like the **North America Nebula**. A vast 50-light-year-wide cloud illumi-nated by nearby stars, it has derived its distinctive shape by being partly obscured by an intervening dark cloud closer to us. Note the greater number of stars in the bright sector of the close-up photo on this page, one indication that this region is farther away than the less star-rich dark area. The North America Nebula is dim but large, filling a good portion of a binocular field. Choose the darkest observing site possible, and use Deneb as a guide.

An even tougher target is the **Veil Nebula**, the tattered remnants of a star that exploded as a supernova approximately 30,000 years ago. Dark skies and experienced eyes are needed to capture it in binoculars or telescopes.

The **Dumbbell Nebula** is a neat puff of star-stuff expanding from a dying star that was once similar to the Sun. Using binoculars, look for a tiny, nearly circular patch. This is one of a class of objects known as planetary nebulas, an unfortunately confusing name that originated with 19th-century observers, who noted a vague resemblance between certain nebulas and the disks of the planets Uranus and Neptune. Remember that the colors of nebulas seen in the time-exposure photos in this book are below the threshold of human vision. These dim objects are always wisps of gray.

The Summer Triangle region is particularly well endowed with **double stars** suitable for binoculars. Double stars are pairs of stars that appear close together, although they are not necessarily physically related. Of the seven listed on facing page, only Epsilon and Zeta Lyrae and 61 Cygni are confirmed binaries, that is, stars which orbit each other. However, all are worthwhile targets for binoculars. Double stars listed as "tight" may require tripod-mounted binoculars. Also, try lying flat on a blanket or an outdoor mattress, and gently rest the rubber eyecups against your cheekbones to steady the view.

THE DOUBLE STAR
Albireo, at the foot of the Northern Cross, is one of the most beautiful stellar pairings in the sky. The brighter star is yellow, its companion blue. A clean split of Albireo usually requires 10-power binoculars or a small telescope.

Veil Nebula

THE FISHHOOK *shape of the constellation Scorpius suggests the tail and stinger of a scorpion. For stargazers, though, the celestial scorpion marks one of the most bountiful regions of the heavens. Under a moonless sky, binocular sweeps through this area bring surprise encounters with glowing fields of stars breached by dark, almost starless chasms. These lanes are actually dark nebulas—galactic gas and dust clouds—closer to us than the starry backdrop, most of which is 5,000 to 8,000 light-years away. In photographs, the Galactic Dark Horse is among the most distinctive of these dark nebulas. The darkest part of the horse, the hind leg and hips (also called Barnard's Pipe Nebula, see page 44), can be clearly seen in binoculars. In the same vicinity, the globular clusters M19 and M62 look like out-of-focus stars. For observers in the northern United States and Canada, however, the lower section of the fishhook is close to the horizon, where atmospheric haze makes viewing difficult.*

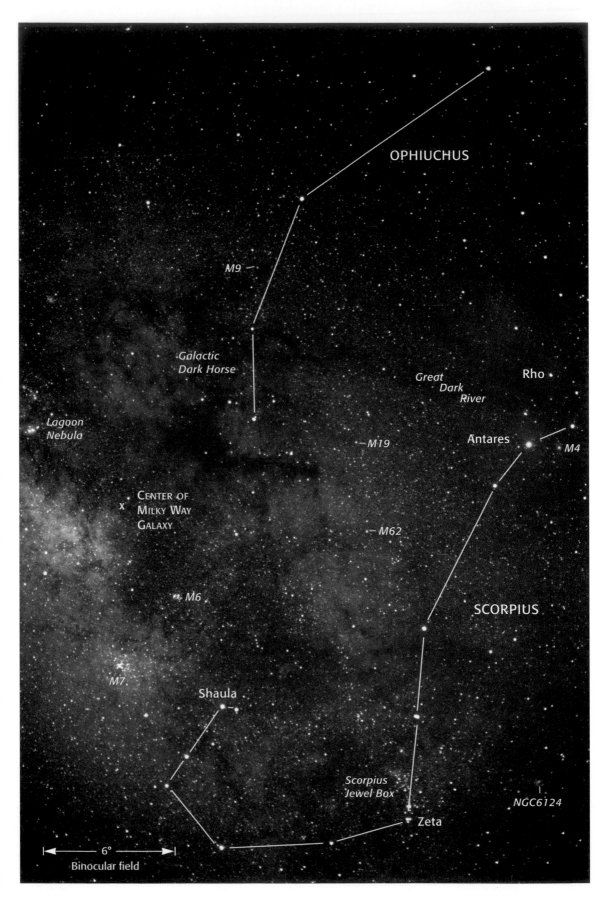

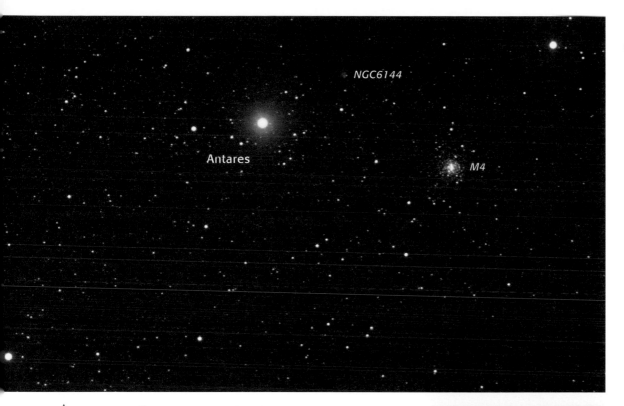

NGC6144

Antares

M4

ANTARES,
the heart of the celestial scorpion, is a red supergiant star about 250 times the diameter of the Sun and 525 light-years distant. The name Antares is ancient Greek and roughly translates to "the rival of Mars." Although not actually red, Antares does have a distinct yellow-orange hue visible to the unaided eye and enhanced by binoculars.

Ancient collections of hundreds of thousands of stars, globular clusters date to the origin of the Milky Way Galaxy. Globulars, compacted by the mutual gravity of their stars into spherical blobs less than 100 light-years across, are spectacular sights in large telescopes. For binocular skywatchers, the easiest to locate of the 140 known globulars is **M4**, just 1.2 degrees to the right of Antares, the lead star in Scorpius. The close-up view of the M4 region above shows Antares and its distinctive flanking stars. Like all globulars, M4 appears as a

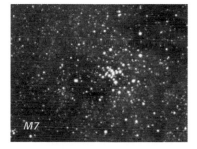

M7

round, fuzzy patch in binoculars. The individual cluster stars in the photo require a 4-inch or larger telescope. At a distance of 6,800 light-years, M4 is the nearest of the globu-

lar clusters. NGC6144, the more distant globular identified in the photo, is too faint to be seen in binoculars.

Just above the star Shaula that marks the scorpion's stinger, look for the clusters **M6** and **M7**, two impressive stellar splashes set in one of the richest Milky Way star fields. These clusters never

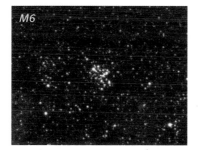

M6

fail to impress. Even the smallest binoculars will show them. Faintly visible to the naked eye, the two clusters have been noted as bright clumps in the Milky Way since Roman times.

Just as impressive in binoculars as the M6-M7 region is the **Scorpius Jewel Box**, containing several overlapping star clusters. Unfortunately, this part of Scorpius is very close to the southern horizon from the northern United States and

Scorpius Jewel Box

NGC6231

Zeta 2   Zeta 1

Canada, and horizon haze interferes with the view.

Just above Antares, the star marked **Rho** is Rho Ophiuchi, which is revealed as a beautiful triple star in binoculars.

Close-Up: Galactic Hub Region

THICKETS OF STARS
so dense that in places, they merge into a golden gauze dominate this view toward the central core of our home galaxy, the Milky Way. The tattered and irregular appearance of the star clouds is caused by intervening dark material—celestial gas and dust—that actually blocks most of our view of the galaxy's bulging hub. A wide river of dark clouds extends from top to bottom in this view. The pinkish blobs mark regions where the clouds are illuminated by pockets of star formation. The brightest of these, the Lagoon Nebula, is shown on facing page. Elsewhere, clumps of stars, such as M7 and M25, are nests of recently created stars that are easy binocular targets. M22, one of the brightest globular clusters, is a distinct glowing sphere in binoculars.

Eagle Nebula
(M16)

Swan Nebula
(M17)

M25

Sagittarius
Star Cloud
(M24)

M22

M23

M21

Trifid Nebula
(M20)

Lagoon Nebula
(M8)

Barnard's Pipe
Nebula

X
CENTER OF
MILKY WAY GALAXY

M6

M7

6°
Binocular field

Sagittarius Star Cloud

View Toward the Center of the Galaxy

Lagoon Nebula

is located in one of the galaxy's spiral arms, 13,000 light-years distant, about halfway to the center of the Milky Way Galaxy. The Star Cloud is approximately 600 light-years long. Above the Star Cloud, the Swan Nebula is a dim gray patch in binoculars and small telescopes.

The **Lagoon Nebula** is a colossal star factory, a giant 50-light-year-wide interstellar cloud being illuminated by newborn stars within. At an estimated distance of 5,200 light-years, the Lagoon is one of only two star-forming nebulas faintly visible to the naked eye from mid-northern latitudes (the other is the famous Orion Nebula, seen in winter). In binoculars, the Lagoon is a distinct oval cloudlike patch with a definite core, like a pale celestial flower. The nebula has a delicate star cluster superimposed on it, making this one of the leading celestial sights of summer night skies.

Like many nebulas, the Lagoon appears pink in time-exposure color photos but is gray to the eye peering through binoculars or a telescope. Human vision has poor color sensitivity at low light levels.

Although our **view toward the center of the galaxy** is heavily obscured by dark gas and dust clouds, we do get a peek at the section shown in detail on this page. The stars here are 24,000 light-years away and distinctly yellow, because the galaxy's core region contains old stars, mostly of this color.

V isible to the naked eye as a bright oval portion of the Milky Way, the **Sagittarius Star Cloud** is a gloriously rich profusion of stellar points in binoculars. This is the densest concentration of individual stars visible in binoculars—roughly 1,000 fill a single field of view. The Sagittarius Star Cloud

OUR GALACTIC HOME

Illustration of the Milky Way Galaxy's structure (below) reveals the location of our Sun, 24,000 light-years from the core. Our view toward the nucleus must penetrate several layers of dark nebulas that hide much of the star-thronged hub. This is why the region seen on facing page appears so heavily rifted by dark rivers among the star fields.

SUN

Close-Up: Celestial Treasures

Andromeda Galaxy (M31)

ANDROMEDA GALAXY, *above, is the nearest galaxy similar to the Milky Way and, at 2.3 million light-years, the most distant object visible to the naked eye. In binoculars, the oval shape is distinct and the two small companion galaxies are faintly seen.*

Pleiades Cluster

Sombrero Galaxy (M104)

At a distance of 2.3 million light-years, the **Andromeda Galaxy** is the most remote object clearly visible to the unaided eye (guide chart, page 38). Gazing at the ghostly oval of the big galaxy in binoculars is one of the great mind-stretching vistas in nature. Long-exposure photographs of the Andromeda Galaxy, such as the one-hour exposure reproduced here, taken through a 6-inch refractor telescope, reveal vastly more detail than the eye can detect, even with a large telescope. Much farther away than Andromeda, at a staggering 50 million light-years, is the giant **Sombrero Galaxy**, a huge edge-on system that marks the practical limit to

DECODING THE JARGON

Galaxies are the largest discrete collections of stars in the universe. The Milky Way Galaxy and the Andromeda Galaxy are classified as spiral galaxies. Other categories are elliptical and irregular.

Globular clusters, like M13 (right), are compact groupings of up to several million stars. More than 140 are associated with the Milky Way Galaxy and trace vast looping orbits around the galactic hub.

Clusters, sometimes called open clusters or galactic clusters, are congregations of a few hundred to a few thousand stars inside our galaxy. The brightest cluster is the Pleiades (facing page), at a distance of 450 light-years. One of the richest is M11 (below).

Nebulas are huge cosmic clouds of gas and dust within our galaxy. They come in two types: bright and dark. The bright ones are often the site of star formation, such as the Lagoon Nebula (page 45), though some, such as the Veil (page 41), are remnants of star death.

Designations of galaxies, clusters and nebulas, such as M31, NGC6633, Stock 2 and IC4756, are simply catalog numbers and, unfortunately, do not indicate object types or classifications.

Hercules Cluster (M13)

HERCULES CLUSTER, *the most famous example of a globular cluster, is 21,000 light-years from Earth. This is a telescopic view. In binoculars, the cluster is a tiny, fuzzy patch in the Keystone of Hercules (see Charts 4, 5 and 6).*

its upper flank, seek compact galactic cluster **M11**, which looks like a grainy disk in binoculars. Dozens of individual stars in M11 explode into view in any size telescope. This dense splash of stars is 5,500 light-years away.

About 12 degrees above M11 is the **S-O Double Cluster**, short for Serpens-Ophiuchus Double Cluster, composed of IC4756 and NGC6633. Even though they are bright enough to be visible to the unaided eye as two faint, hazy patches, this duo is not well known to many amateur astronomers. Just three degrees apart, they fit neatly into a binocular field.

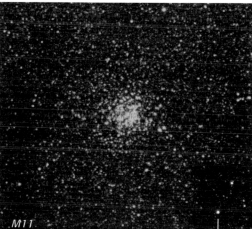

50mm binoculars. Use Chart 3 for guidance, and look for a tiny, elongated smudge. You'll be seeing the combined light of hundreds of billions of stars—light that has taken 50 million years to reach your eyes.

At right is a sector of Chart 6 showing parts of the constellations Aquila, Serpens, Scutum and Ophiuchus. It contains the **Scutum Star Cloud**, a rich section of the Milky Way that resolves into a gorgeous sprinkling of stars in binoculars. On

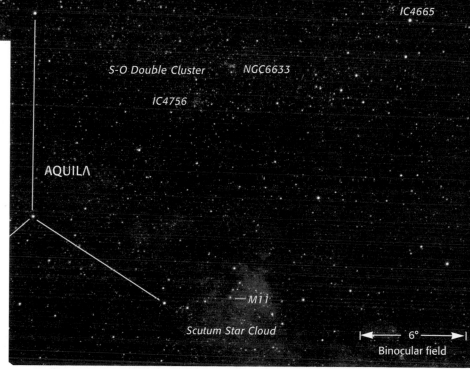

The Sky *in* Other Seasons

Beyond summer stargazing: the rest of the year's constellations

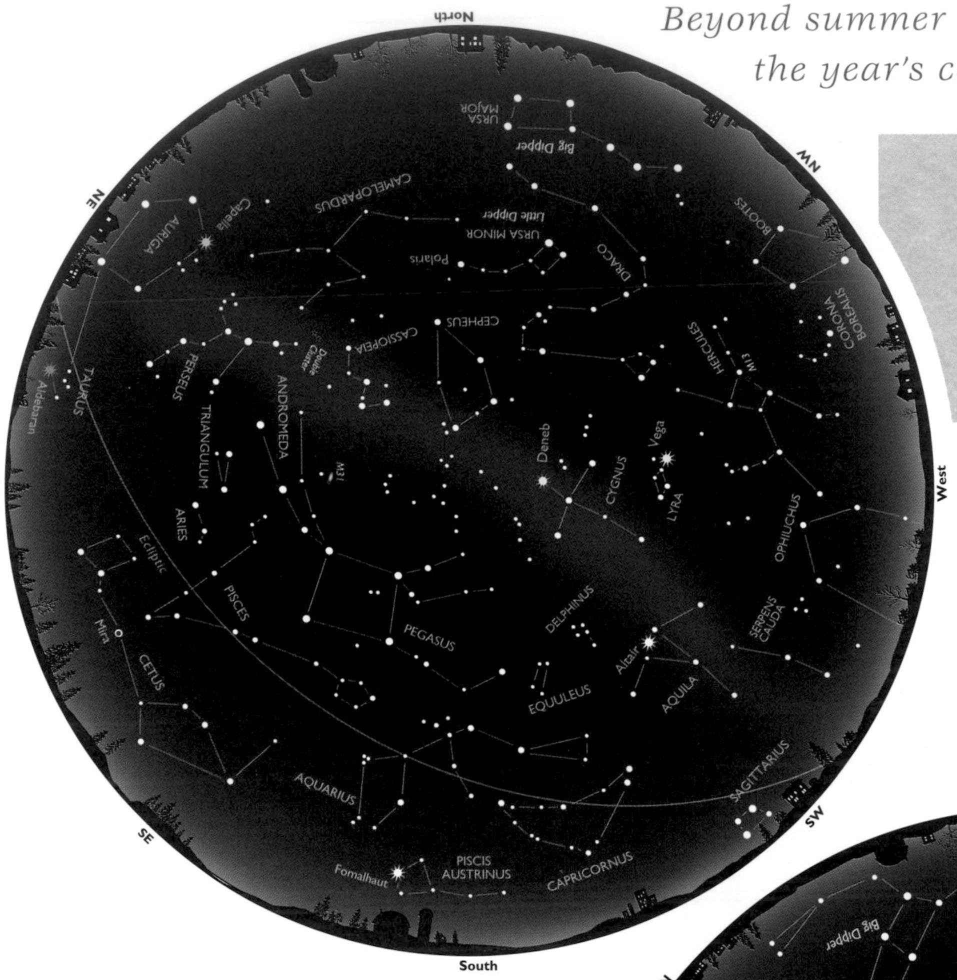

EARLY AUTUMN	
Early Sept.	11 pm-1 am
Late Sept.	10 pm-12 am
Early Oct.	9-11pm
Late Oct.	8-10 pm
Early Nov.	6-8 pm
Late Nov.	dusk-7 pm

LATE AUTUMN	
Late Oct.	11 pm-1 am
Early Nov.	10 pm-12 am
Late Nov.	9-11 pm
Early Dec.	8-10 pm
Late Dec.	7-9 pm
Early Jan.	dusk-8 pm

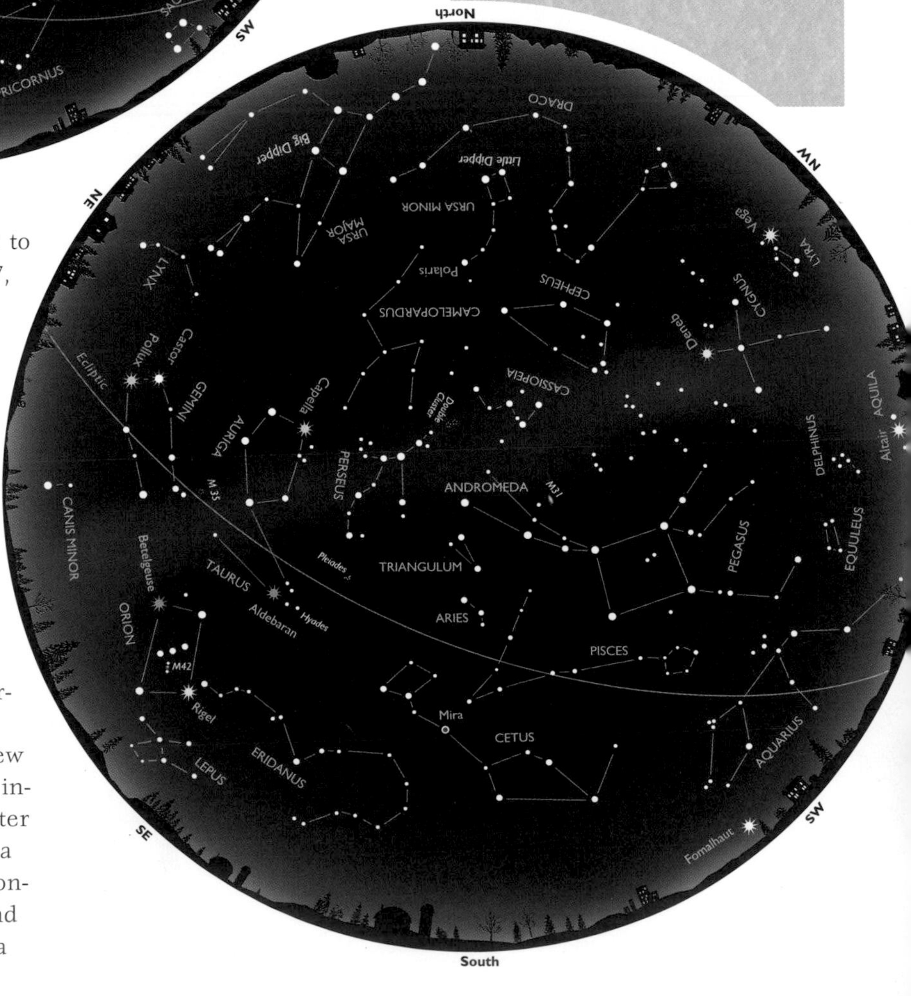

OTHER SEASONS *offer many more constellations to add to your stargazing repertoire. For instructions on how to use the circular star charts, see pages 20-23.*

Our main Charts 1 to 7, on pages 24-37, have already introduced you to some of the stellar tracts seen here. In particular, almost all of the early-autumn sky (above) is given detailed coverage on Charts 6 and 7. But there is a lot of completely new territory to be explored by the enthusiast who might be inspired to indulge in year-round stargazing.

Foremost among the new stellar galleries worthy of investigation is the mid-winter group centered on Orion, a distinctive and majestic constellation that ranks second only to the Big Dipper as a stargazing guidepost. Orion and the surrounding constellations Taurus, Auriga, Gemini, Canis Major and Canis Minor constitute the brightest collection of stars in the entire sky visible from North America. Spread high in the southern sky on January, February and March evenings, this stellar assemblage can give the impression that stars shine brighter in the crisp winter air. However, this is an illusion. It is the greater *number* of bright stars seen in winter that produces this misperception, not the clarity of the air. In most parts of North America, the clearest nights in winter are no clearer than the best nights in other seasons.

In the northern quarter of these charts, the same stars and constellations prevail as those seen on Chart 1 (page 25). These are the circumpolar stars that were mentioned on page 19. They don't rise or set from mid-northern latitudes, but rather, they take a leisurely annual stroll in a counterclockwise pivot, with Polaris at the hub.

ORION IN WINTER, *the king of the constellations, boasts more bright stars than any other stellar configuration.*

WINTER

Late Dec.	11 pm-1 am
Early Jan.	10 pm-12 am
Late Jan.	9-11 pm
Early Feb.	8-10 pm
Late Feb.	7-9 pm
Early Mar.	dusk-8 pm

EARLY SPRING

Late Feb.	11 pm-1 am
Early Mar.	10 pm-12 am
Late Mar.	9-11 pm
Early Apr.	9-11 pm
Late Apr.	8-10 pm

Meteors

A typical meteor is smaller than a peanut

METEOR WATCHING
requires no instrumentation. Binoculars and telescopes, with their narrower sweep of the sky, are less suited to the task of scanning for meteors than are the eyes alone. This makes meteor watching an ideal group activity, where friendly conversation will be interrupted by the "oohs" and "aahs" of delight as meteors etch the darkness. The Perseids (see table on facing page) are by far the best shower for group watching, but only if the Moon is absent. Use reclining lawn chairs for a panoramic view of the sky. If possible, face northeast or east, and select a place that is dark enough for you to see the Milky Way. As a general rule, meteor showers peak after midnight.

At 12:37 a.m., on August 25, 1995, a group of amateur astronomers at the annual Starfest convention near Mount Forest, Ontario, were just packing up their equipment after several hours of telescope observing when they were startled by a dazzling object brighter than the full Moon hurtling across the partly cloudy sky. Lasting for more than 10 seconds as it traced a path from north to south, the object looked like a glowing white-hot ember shedding molten droplets as it passed overhead.

What the astronomy buffs had seen was a classic fireball, a brilliant meteor produced when a chunk of rock from deep space collides with the Earth's atmosphere and burns up. The celestial boulder could have been as large as a shipping trunk, but it was more than likely between a baseball and a basketball in size. The object may have com-

pletely vaporized in the atmosphere, or it may have broken into fragments, some of which might have reached the ground. In this case, no pieces were ever found. (Jargon note: The object seen streaking across the sky is a *meteor*, but a recovered piece is a *meteorite*.)

Fireball meteors like that witnessed by the Starfest group are extremely rare. They are the most dramatic examples of what are commonly called shooting stars, although they are not related to stars in any way. The typical meteor is a piece of cosmic debris, usually smaller than a peanut, that strikes the Earth's upper atmosphere at 50 times the velocity of a rifle bullet and burns up from friction almost instantly. It is this incineration phase and its rapid heating of the air along the trajectory that is seen as a meteor momentarily searing the blackness.

METEORIC ORIGINS

Astronomers have estimated that up to 1,000 tons of meteoric dust and rock enter the Earth's atmosphere every day. Most of it comes from comets.

When a comet approaches the Sun to within the orbit of Mars, its icy surface is vaporized by solar radiation. Dust and debris encased in the ice since the solar system's formation nearly five billion years ago are released to drift into space. Some of this eventually falls to Earth as meteors.

Some large meteors come from the asteroid belt between Mars and Jupiter. These meteors are literally chips off the large rocky objects that orbit in this zone by the thousands.

On a prolonged watch on any given night, you will inevitably see a handful of meteors randomly around the sky. Astronomers call these sporadics. Brighter meteors can leave trails, like jet contrails, though they seldom last for more than a few seconds.

Next to eclipses and bright comets, the celestial events that attract the most publicity are meteor showers. These are predictable annual events when, for one or two nights, the count of meteors jumps to 3 to 10 times the sporadic rate.

The highest-yielding and best known of the annual meteor showers are the Perseids and the Geminids (see table below). At their best, they yield one meteor per minute.

Major meteor showers such as the Perseids are predictable, because comets leave streams of meteoric debris in their wake, like a load of sand leaking from a moving truck. As Earth travels in its orbit around the Sun, it intersects various comet orbits each year at the same place, hence the predictability of prominent meteor displays.

METEOR SHOWERS *occur when Earth passes through debris scattered along the orbit of a comet (see diagram, left). The Perseids, the best-known meteor shower, are sand- and gravel-sized rubble from Comet Swift-Tuttle, last seen in 1992. Perseid meteors appear to radiate from the constellation Perseus. The illustration at the top of this page shows an hour's worth of shower meteors appearing to emerge, warp-drive fashion, from the radiant.*

SUN

EARTH

MAJOR ANNUAL METEOR SHOWERS

Name of Shower	Radiant	Date of Maximum	Hourly Rate at Maximum*
Quadrantid	NE (Draco)	Jan. 3	10-50
Lyrid	NE (Lyra)	Apr. 21	5-15
Eta Aquarid	E (Aquarius)	May 4	10-20
South Delta Aquarid	SE (Aquarius)	Jul. 27-29	10-20
Perseid	NE (Perseus)	Aug. 12	30-70
Orionid	E (Orion)	Oct. 20	10-30
Leonid	E (Leo)	Nov. 17	10-20
Geminid	E (Gemini)	Dec. 13	30-80

*The range of values reflects the variations in the strength of the meteor displays from year to year. These figures do not include the half-dozen or so sporadic meteors seen each hour.

Auroras

During sunspot maximum, Canada and the northern United States are treated to dazzling displays of northern lights

They appear most often as delicate, pale green curtains shimmering in the northern sky, swaying and billowing as if being caressed by some cosmic wind. Occasionally, the sky erupts into a kaleidoscope of pulsating shafts and arches that fill the heavens like the vaultings of an immense, ethereal cathedral. The native peoples of Siberia, Lapland and Canada's West Coast thought that such displays were the spirits of great warriors battling in the sky, but we know them today as the aurora borealis, or northern lights.

Auroras range from a pale greenish white glow near the northern horizon to intense red, green and purple spears and ribbons that fill the sky, magically floating among the stars. The typical aurora is located a few hundred kilometers up, in the outer reaches of the Earth's atmosphere, but its origin can be traced to the Sun. Eruptions on the Sun's surface, called solar flares, liberate vast amounts of electrically charged particles into space. The charged particles —actually just parts of otherwise neutral atoms—reach Earth, follow our planet's natural

magnetic field and are funneled into a continent-sized ring around the magnetic north pole in Canada's Arctic (a similar ring occurs over Antarctica).

The interaction between the incoming particles and atoms and air molecules in the atmosphere releases energy as light. The Earth's upper atmosphere effectively acts like a television screen, glowing when it is bombarded by the electrically charged particles. This is happening all the time, but it is often dim, or the auroral ring is too far north

to be visible from the more populated parts of Canada and the northern United States.

When solar activity increases and more particles arrive at Earth, the aurora brightens, the ring expands to the south and the display becomes potentially visible to millions.

Although auroras can happen anytime, the majority oc-

BRILLIANT AURORAS, *like the ones shown here, were seen during the last sunspot maximum in 1989-91. More are expected during the next maximum in 2000-02. When seen at their brightest, auroras display distinct colors, with green and red predominating. The photo above and the city image at far right (facing page) were taken in central Alberta. All other photos were taken at 44 degrees north latitude, near Lake Ontario.*

cur during the two to three years when sunspots are at their peak and solar flares are most common. The last sunspot maximum was 1989-91, and the next should be 2000-02. Historically, more intense auroras seem to come during the months of March, April, September and October, but any advance predictions are impossible, because major displays are driven by unpredictable storms on the Sun.

One of the biggest solar eruptions on record produced a sensational aurora on the night of March 12-13, 1989. Seen as far south as Guatemala, the light show was the subject of front-page newspaper stories in Florida, Texas and parts of Europe. So brilliant was the display that it even overpowered the artificial nighttime illumination in many large cities. The aurora was the most impressive since the sunspot maximum of the late 1950s. The auroral light was almost as bright as the full Moon, enveloping the entire sky in shimmering, multicolored rays that fanned out from the zenith like the ribbing of an umbrella. Veteran stargazers ranked it second only to a total eclipse of the Sun. Other spectacular displays occurred on the nights of March 20, 1990, and November 8, 1991.

Reports from people who say they can hear auroras are so persistent that the phenomenon must almost certainly be real, despite science's inability to explain it. There are hundreds of accounts of crackling, swishing and hissing sounds associated with auroras, but only some people seem to be able to hear them.

Lunar Eclipses

About once a year, a partial or total eclipse of the Moon comes to your neighborhood

Most wall calendars display the Moon's phases, but few note the most dramatic lunar phenomenon: a total eclipse of the Moon. A total lunar eclipse occurs when the Moon is completely within the Earth's shadow. The Moon is dimmed but usually does not disappear, due to a small amount of sunlight that leaks through the Earth's atmosphere. The best part is that the whole event is easily visible to the naked eye. If you can see the Moon on eclipse night, you can see the eclipse.

A surprising aspect of a lunar eclipse for many people is the color of the Moon when it is dunked into the Earth's shadow. Because red and orange wavelengths of light leak into the shadow more readily than do other colors, the color of the eclipsed Moon can range from golden buff to rusty brown to burnt toast. The shading comes as sunlight is lensed into the shadow by the Earth's atmosphere.

To visualize the geometry of a lunar eclipse, picture the sun shining on Earth, which in turn casts a slender cone-shaped shadow 1.3 million kilometers (800,000 mi) into space. At the Moon's distance, 384,000 kilometers (240,000 mi), the cone is about $2^1/_2$ times the lunar diameter. Lunar eclipses can occur only at full Moon, when the Moon is on the side of Earth directly opposite the Sun. Although

THE EARTH'S SHADOW *slowly sweeps over the Moon during a partial lunar eclipse (above). The unusually dark total lunar eclipse of December 9, 1992 (center), reduced the full Moon's brightness by a factor of one million, making it no more luminous than a bright star. Binoculars are ideal tools for lunar-eclipse watching.*

FUTURE LUNAR ECLIPSES VISIBLE FROM NORTH AMERICA

Times given are for mid-eclipse; begin watching at least one hour earlier to see complete event; convert to your time zone if necessary.

1996, September 26: best total lunar eclipse of the 1990s; visible over entire continent except extreme NW (10:54 pm, EDT)
1997, March 23: partial (90%); visible over entire continent (11:39 pm, EST)
1999, July 28: partial; visible only from western third of continent (4:33 am, PDT)
2000, January 20: total; visible over entire continent; along with total eclipse of 2004, this is the best of the decade (11:43 pm, EST)
2000, July 16: total only for Hawaii; seen as partial from West Coast (6:55 am, PDT)
2001, January 9: total; visible only as partial at dusk from New England and Atlantic Canada (3:20 pm, EST)
2003, May 15: total; visible over entire continent except extreme NW (11:40 pm, EDT)
2003, November 8: total; visible over entire continent except West Coast (8:19 pm, EST)
2004, October 27: total; visible over entire continent (11:04 pm, EDT)
2007, March 3: total for Great Lakes and eastward; partial for rest of continent except extreme NW (6:20 pm, EST)
2007, August 28: total for Rocky Mountain region and westward; partial for rest of continent except extreme NE (3:37 am, PDT)
2008, February 20: total; visible over entire continent (10:26 pm, EST)
2010, December 21: total; visible over entire continent (3:17 am, EST)

the Moon orbits Earth every $29^1/_2$ days, there are usually only two lunar eclipses a year. Since the Moon's orbit is tipped five degrees from the plane of the Earth's orbit, it is only where these planes intersect that eclipses can occur. Given the inevitable cloud-outs, unfavorable timing (the eclipse may happen while the Moon is

there is a wonderful element of the unknown, because the color and brightness of the Moon depend on the clarity of the atmosphere and the density of the cloud cover around the Earth's circumference. More clouds block more light and darken the eclipse. Exceptionally dark lunar eclipses are caused by the added obscuring effects of volcanic gas and ash in the upper atmosphere, as was the case during the December 9, 1992, lunar eclipse (left), when the Moon almost disappeared as a result of the lingering effects of the Mount Pinatubo eruption in June 1991.

ECLIPSE PHOTOGRAPHY

Lunar eclipses can be photographed with both video and film cameras, provided they are secured on a solid camera tripod. For video, practice beforehand by taping the Moon using various camera settings at different lunar phases. This will give a range of conditions that, to some degree, simulates an eclipse.

With 35mm cameras, telephoto lenses of at least 200mm focal length are recommended to show detail on the Moon. Shorter lenses can be used to set the scene, as was done with a 50mm lens in the photo above. The three

large lunar-eclipse images were taken with a 5-inch telescope. Lunar eclipses vary enormously in brightness, so exposures should be generously bracketed.

The huge difference in brightness between the still fully sunlit portion and the portion in the shadow is evident in the photo below. In the bottom photo, the Moon is immersed in the shadow. Binoculars will reveal all the detail seen here. Mark your calendar for the next lunar eclipse!

COLORFUL SHADING *distinguished the total lunar eclipse of August 16, 1989 (both photos below), which was widely observed throughout North America. Upper image shows the Moon 80 percent inside the Earth's shadow.*

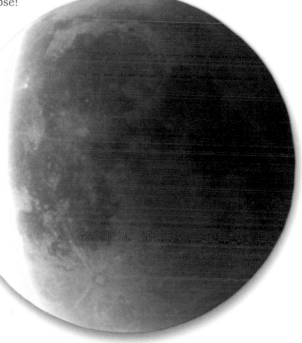

not in view from your location) and the fact that roughly half of all lunar eclipses are partial and far less impressive, the odds of seeing a total lunar eclipse more than twice a decade are relatively slim.

OBSERVING LUNAR ECLIPSES

No equipment is needed to see a lunar eclipse, but I find binoculars and a padded lawn chair are my best accessories. Binoculars enhance the eclipse's subtle colors and shadings just as well as a telescope. A typical total lunar eclipse lasts three hours or more—one hour each for the entering and exiting partial phases and up to 90 minutes for totality.

Although astronomers know *when* everything will happen during an eclipse,

The Sun *and* Solar Eclipses

With proper filters and precautions, you can examine the nearest star

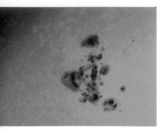

A SUNSPOT GROUP *larger than Earth (above) is not an uncommon occurrence on the face of our nearby star. A properly filtered small telescope can reveal this much detail. Multiple exposure (right) shows the Moon progressively moving in front of the Sun during the partial eclipse that precedes totality. Telescopic view of setting Sun (below) is distorted by refraction from the Earth's atmosphere.*

stronomy is not entirely a nighttime pursuit. The nearest star, our Sun, with its ever-changing sunspot blemishes, provides a unique daytime target for stargazing enthusiasts. The Sun is also the central player in the most spectacular of all celestial phenomena: a total solar eclipse.

First, let's consider the Sun on a typical day. It's too bright to look at without protection, and it's downright dangerous to examine with unfiltered binoculars or telescopes. A quick glance directly at the Sun through such equipment can cause serious eye damage.

The only completely safe solar filters for telescopes are sold by telescope-supply companies and are designed to fit in front of the optical system, like the one affixed to the telescope at far right. Screw-in eyepiece filters (facing page, center), often sup-

plied with the so-called beginners' telescopes that I critiqued on page 8, are *not* safe. After a few minutes of absorbing the intense concentrated heat at the telescope's focus, eyepiece filters inevitably crack and immediately allow a shaft of unobstructed sunlight to reach the eye. If you

have one of these bogus filters, throw it away.

Two inexpensive filter materials are especially suited to solar viewing with both the naked eye and binoculars. One is welder's filter plate grade 14, available in 2-by-4-inch rectangles at many welder's supply shops. Securely tape one of these filters over each of the front lenses of your binoculars.

The second inexpensive filter material is tough Mylar, with a highly reflective metal coating. This thin, flexible material is sold in frames as "eclipse glasses" and is also available in loose sheets from telescope dealers.

Filtered instruments are tough to aim at the Sun, because the field of view appears totally black, except for the solar disk. The best plan is to mount filtered binoculars on a tripod and align them by watching the shadow of the binoculars on the ground. This will get you close. The same shadow-aiming technique works with filtered telescopes.

A method of solar viewing that requires no special filter utilizes a telescope or binoculars as a projection lens (below). An accessory projection screen works well, but a hand-held white card will do. Although this method is good for group instruction, more detail in the sunspots is seen by filtered viewing, as described above. To avoid overheating and damaging larger optical systems used for projection, limit the intensity by taping a card, with a 50-to-60-millimeter hole cut

FUTURE TOTAL SOLAR ECLIPSES

Paths of totality are typically 100 kilometers (60 mi) wide but can stretch across a continent; for maps and other details, consult astronomy magazines up to 12 months before eclipse day.

1997, March 9: centerline of totality extends over an inaccessible sector of Asia, from Mongolia into Siberia

1998, February 26: totality touches northern Colombia and Venezuela and the Caribbean islands of Antigua and Montserrat

1999, August 11: totality runs through SW England, northern France, Germany, Austria, Hungary and Romania

2001, June 21: South Atlantic and south-central Africa

2002, December 4: Mozambique and South Pacific

2005, April 8: South Pacific

2006, March 29: north-central Africa and Turkey

2008, August 1: Arctic and Siberia

2009, July 22: China and equatorial Pacific

The next total eclipse of the Sun over the United States or Canada is a long way off—August 21, 2017—when the path of totality will stretch from Oregon to central Nebraska, then on through Tennessee and South Carolina. Less than seven years later, on April 8, 2024, another cross-continent total solar eclipse occurs, this one running through central Texas, Arkansas, Indiana, Ohio, southern Ontario and southern Quebec.

PARTIAL SOLAR ECLIPSES VISIBLE FROM NORTH AMERICA

Times when partial eclipses occur can vary by several hours, depending on the observer's location; check almanacs or astronomy magazines for details.

1996, October 12: visible in Atlantic Canada north of a line from Halifax, Nova Scotia, to Fredericton, New Brunswick (about 10% of the Sun will be covered)

1998, February 26: visible from east and south of a line from Los Angeles to Milwaukee (20% in New York; 45% in Miami)

2000, December 25: northeastern North America (60% in Great Lakes area around 2:35 pm, EST)

2001, December 14: western North America

2002, June 10: western North America

2005, April 8: south of a line from San Diego to Philadelphia

in it, to the front of the instrument.

SOLAR ECLIPSES

If you ever have an opportunity to stand in the Moon's shadow and witness a total eclipse of the Sun, grab it. The awesome spectacle is worth a trip to another continent—as will be necessary for some time to come.

Group excursions to view total eclipses are often organized by astronomy clubs and the major astronomy magazines. These usually include seminars by eclipse experts and tours of astronomical sites in the locality.

The more common and far less impressive partial solar eclipse can be observed with the same filter techniques used for solar observing.

SOLAR FILTERS
that fit over the front of a telescope (above) are recommended, rather than eyepiece filters (top left).

57

Comets

Ethereal and elusive, comets are icy visitors from afar

HALLEY'S COMET *made its last visit to the Earth's vicinity in 1986 (above and facing page, far right). It will return in 2061.*

Comets are the phantoms of the solar system. They appear as if from nowhere, sprout ghostly tails of gas and dust as they sweep closer to the Sun, then retreat into the abyss and disappear, often never to be seen again. A bright comet, pointing toward the horizon like a sword or the finger of doom, is unlike anything else in the night sky. Our ancestors, understandably, regarded comets as bad omens.

Anyone who has seen a comet only in photographs might think that these objects are dashing through the Earth's atmosphere, to be consumed by friction like meteors. Not so. These are huge objects, typically as far away as Mars or Venus, with tails millions of kilometers long. They can be visible for months, cruising through the constellations.

WHAT ARE COMETS?

Week-old snow piled at the sides of roads in mid-winter has a lot in common with comets. The main comet ingredient is frozen water, with cosmic dust mixed in. A small percentage of the ice is frozen ammonia, methane and carbon dioxide, but if you had a handful of the stuff, it would easily substitute for darkened, dirt-crusted roadside snow.

A comet is a chunk left over from the formation of the giant planets (Jupiter, Saturn, Uranus

FAMOUS AND INFAMOUS COMETS

Descriptions of comets of the past, particularly from the 19th century, give the impression that dazzling comets with enormous sweeping tails were commonplace. Much of this has to be attributed to overenthusiasm —some would say exaggeration— among observers of the time. For instance, the painting of the Great Comet of 1882 (left) shows a telescopic view of the comet misleadingly superimposed over a pastoral observatory scene.

But there is no denying that some comets have put on magnificent displays. Astronomers estimate that Halley's Comet in the year 837 reached the brightness of a thin crescent Moon, which likely sets the record for the past 2,000 years. In April 1910, Halley was again very impressive as it reached first magnitude, the brightness of stars such as Deneb and Altair. More recently, Comet West offered a similarly dramatic display in 1976. Other bright comets of the 20th century include Comet Bennett of 1970, Comet Ikeya-Seki of 1965, Comet Mrkos and Comet Arend-Roland, both seen in 1957, and the Great Comet of 1910, which graced the skies four months before Halley.

On the other hand, some comets that were predicted to be spectacular fizzled out. The most notorious of these was Comet Kohoutek, which was less than one-tenth of 1 percent as bright as astronomers expected when it was at its best in January 1974. Similarly, Comet Austin, hailed in advance as a "monster comet" by the leading astronomy magazines, was barely visible to the naked eye when it reached its peak in April 1989.

But the comet that many people remember as a major disappointment was Halley in 1986. This time, the astronomers were right. Because Comet Halley was more distant than it had been in 1910, they had predicted that it would be far less impressive. But romanced by the famous comet's reputation and by a surge of media hype, many novice observers expected fireworks and, instead, were served a mere celestial smudge.

and Neptune) and is, on average, about the size of one of the Rocky Mountains. Astronomers estimate that there are at least a trillion comets located at the rim of the solar system, reaching perhaps a quarter of the way to the nearest star.

The vast majority of comets never come closer than Neptune, 30 times the Earth's distance from the Sun. But a few renegades venture inside the orbit of Mars. Once a comet comes that close to the Sun, its surface is vaporized by solar radiation. Along with the release of gas, embedded dust is expelled, forming a gas/dust cloud surrounding the frozen snowball. That cloud, in turn,

is swept back into the classic comet tail by the pressure of sunlight and the force of the solar wind (electrically charged particles from the Sun). The result: one of the most treasured shows in the night sky.

COMET WATCHING

Most comets are very faint and of interest only to researchers and comet buffs. About once a year, though, a comet becomes bright enough to breach the limit of naked-eye visibility. At such times, binoculars might show a defi-

nite tail, and you may want to hunt the comet down.

Roughly once in a generation, a really bright comet arrives. When it does, don't spare any effort to see it.

Finding out about the comet is the first task. The astronomy magazines listed on page 62 will inform you of current comet activity. There are also comet "hot lines" on the Internet, if you are so equipped. Your local astronomy club is another good source of information.

Keep in mind that comets are very susceptible to light

pollution and should be observed from as dark a location as possible, well away from interfering city lights.

Know your directions and constellations. If the comet is rising low in the northeast near Andromeda just before dawn, you want to be able to find it. Most comets are subtle and require searching.

Use averted vision. With the comet centered in binoculars, look off toward the edge of the field of view, but keep concentrating on the comet with your peripheral vision. Human eyesight is more sensitive to dim light when it's off the central axis of vision. The comet should suddenly appear brighter, with a more distinct and possibly longer tail.

Planet Visibility 1996-2010

THE PLANETS: NAKED-EYE APPEARANCE

Venus, sometimes called the morning or evening star, is the brightest planet, easily recognized by its dazzling diamond-white appearance and substantially greater brilliance than any other starlike object.

Jupiter is the second brightest planet, but it still outshines all the fixed stars. It shines with a slightly off-white creamy hue. Because it takes 12 years to orbit the Sun, Jupiter spends about a year in each constellation of the zodiac.

Saturn, similar in color to Jupiter, is the same brightness as the brightest stars. The slowest-moving naked-eye planet, Saturn takes 29.5 years to orbit the Sun and spends about two years in each constellation of the zodiac.

Mars shines with a definite ocher or pale rusty hue. It varies greatly in brilliance, ranging from the brightness of the Big Dipper stars to that of Jupiter. At times, it moves rapidly across the heavens, spending as little as one month in a zodiac constellation.

Mercury is the most elusive of the naked-eye planets, being visible only near the western horizon at dusk or the eastern horizon at dawn for two-week intervals a couple of times during the year.

PROMINENT PLANETARY CONJUNCTIONS 1996-2010

The Moon and the brighter planets are the night sky's most conspicuous celestial objects, especially when they have close encounters with each other—events called conjunctions. The tabulation below includes the best, though far from all, of the conjunctions visible from North America through 2010. Starred items are either particularly impressive or relatively rare. Note especially the April/May 2002 event.

March 26, 1998*	Thin crescent Moon passes directly in front of Jupiter (called an occultation), 5-6 am, EST; visible only in eastern North America
April 23, 1998*	Venus less than 0.5° from Jupiter in morning sky
February 23, 1999*	Venus 0.3° from Jupiter in evening sky; spectacular
March 19, 1999	Venus 2° from Saturn; crescent Moon nearby
February 2, 2000	Crescent Moon 1° from Venus in morning twilight
early April 2000	Jupiter, Saturn and Mars cluster low in evening sky; crescent Moon joins them Apr. 6
July 15, 2001	Venus and Saturn less than 1° apart in early-morning sky
July 17, 2001	Crescent Moon, Venus and Saturn form a 3° triangle in morning twilight
August 6, 2001	Venus 1.3° from Jupiter in early-morning sky
November 6, 2001	Venus 0.8° from Mercury in morning twilight
February 22, 2002	Gibbous Moon 0.3° from Jupiter around midnight
April 22 - May 13, 2002*	For the first time in more than a generation (since May 1980), all five naked-eye planets are lined up above the western horizon at dusk; crescent Moon joins the group on May 13; best alignment will be late April, with planets in this order from horizon up: Mercury, Venus, Mars, Saturn and Jupiter; star Aldebaran near Saturn
May 14, 2002	Crescent Moon 1° from Venus low in west at dusk
June 3, 2002	Venus and Jupiter 2° apart low in west at dusk
June 14, 2002	Crescent Moon less than 2° from Venus (3° in western North America)
November 5, 2004*	Venus and Jupiter 0.6° apart in the east in morning twilight
December 7, 2004*	Crescent Moon passes directly in front of Jupiter from about 4-5 am, EST (not visible from western half of North America)
June 25, 2005	Venus, Mercury and Saturn cluster within 1.5° of each other in west at dusk
June 27, 2005*	Venus just 0.1° from Mercury in west at dusk; closest encounter of these two planets visible from North America since 1965
June 17, 2006	Saturn and Mars 0.5° apart and just 1° from Beehive star cluster in evening sky
May 19, 2007*	Crescent Moon just 1° from Venus in evening sky; *very* impressive
July 1, 2007	Venus 0.8° from Saturn low in west in evening twilight
February 1, 2008	Venus and Jupiter 0.6° apart in east in morning twilight
February 4, 2008	Venus, Jupiter and crescent Moon cluster in morning sky before dawn
December 1, 2008*	Venus, Jupiter and crescent Moon cluster in a 3° triangle at dusk
December 31, 2008	Jupiter and Mercury 1.2° apart low in west at dusk
February 27, 2009	Crescent Moon less than 2° from Venus in evening sky
October 13, 2009	Venus and Saturn 0.5° apart in early-morning sky

WHERE TO FIND JUPITER AND SATURN 1996-2010

	Jupiter	Saturn
1996	In Sagittarius all year; Sun's glare interferes in early Jan.	In Aquarius to Feb.; too close to Sun to mid-Apr.; in Pisces rest of year
1997	Buried in twilight glow Jan. and Feb.; in Capricornus rest of year	In Pisces all year but lost in solar glare Mar. and Apr.
1998	In Aquarius in Jan.; too close to Sun Feb. and Mar.; rest of year on Aquarius-Pisces border	In Pisces until late Mar., when twilight interferes; in Aries after late May
1999	In Pisces Jan. and Feb.; near Sun mid-Mar. to mid-May; in Aries remainder of year	In Aries all year; Sun's glare rules out observation Apr. to early June
2000	In Aries to mid-Apr.; Sun's glare prevents observation to late June; in Taurus thereafter	Jupiter and Saturn less than 15° from each other all year
2001	In Taurus until solar glare scoops it up in May; seen in Gemini in morning sky after late July	Saturn less than 15° to right of Jupiter until solar glare interferes late Apr. to early July; then look near Aldebaran
2002	In Gemini until solar glare interferes in early June; seen in Cancer from mid-Aug. to year's end	Near Taurus-Gemini border all year; solar glare prevents viewing from early May to early July
2003	In Cancer to mid-July; emerges from solar glare in Leo in mid-Sept. and remains there into 2004	Visible in Gemini except for late May to late July, when lost in twilight glow
2004	Look for Jupiter in Leo until mid-Aug., when it is lost in twilight; in Virgo in morning sky after mid-Oct.	Remains in Gemini; not visible from early June to early Aug. because of solar glare
2005	In Virgo until mid-Sept., when lost in twilight; in Libra after mid-Nov.	In Gemini until twilight interferes in mid-June; in Cancer after late Aug.
2006	Near Virgo-Libra border for entire year except after mid-Oct., when Sun's glare prevents convenient viewing	In Cancer until lost in solar glare from early July to early Sept.; in Leo thereafter
2007	In Scorpius all year; too close to Sun for observing after mid-Nov.	In Leo to right of Regulus until twilight interferes from mid-July to mid-Sept.; then look to left of Regulus
2008	In Sagittarius all year; Sun's glare prevents viewing in Jan.	In Leo all year; Sun's glow interferes from late July to early Oct.
2009	Lost in twilight glow through to early Mar.; in Capricornus rest of year	Near Virgo-Leo border; solar glare interferes from mid-Aug. to mid-Oct.
2010	In Capricornus in Jan.; twilight interferes with viewing through to early Apr.; in Aquarius thereafter	In Virgo all year; twilight glow prevents viewing from late Aug. to late Oct.

WHERE TO FIND MARS 1996-2010

	Jan.	Feb.	Mar.	Apr.	May	Jun.	Jul.	Aug.	Sep.	Oct.	Nov.	Dec.
1996	—	—	—	—	—	Tau	Tau	Gem	Cnc	Leo	Leo	Leo
1997	Vir	Vir	Vir	Leo	Leo	Vir	Vir	Vir	Lib	Sco	Sgr	Sgr
1998	Cap	—	—	—	—	—	—	—	Leo	Leo	Vir	Vir
1999	Vir	Vir	Lib	Lib	Vir	Vir	Vir	Lib	Sco	Sgr	Sgr	Cap
2000	Aqr	Psc	Psc	Ari		—	—		—	Leo	Vir	Vir
2001	Lib	Sco	Sco	Sgr	Sco	Sco	Sco	Sco	Sgr	Sgr	Cap	Aqr
2002	Psc	Psc	Ari	—	—	—	—	—	—	—	—	—
2003	Lib	Sco	Sgr	Sgr	Cap	Aqr	Aqr	Aqr	Aqr	Aqr	Aqr	Aqr
2004	Psc	Psc	Ari	Tau	Tau	—	—	—	—	—	—	—
2005	—	—	—	Cap	Aqr	Psc	Psc	Ari	Ari	Ari	Ari	Ari
2006	Ari	Tau	Tau	Gem	Gem	—	—	—	—	—	—	—
2007	—	—	—	Aqr	Psc	Psc	Ari	Tau	Tau	Gem	Gem	Gem
2008	Gem	Tau	Tau	Gem	Gem	Cnc	Cnc	Leo	—	—	—	—
2009	—	—	—	—	—	Ari	Tau	Tau	Gem	Gem	Cnc	Cnc
2010	Cnc	Cnc	Cnc	Cnc	Cnc	Leo	Leo	Vir	Vir	—	—	—

Dashes indicate that the planet is in twilight glow and difficult to observe. Three-letter abbreviations are the 12 zodiac constellations. Use all-sky charts on pages 20-23 and 48-49 to determine where Mars is seen in the sky for a specific month (i.e., the constellation it is passing through).

BEST TIMES FOR SIGHTING MERCURY 1996-2010

1996	mid- to late Apr.	early to mid-Oct.
1997	late Mar. and early Apr.	mid- to late Sept.
1998	early to mid-Mar.	late Aug. and early Sept.
1999	late Feb. and early Mar.	mid- to late Aug.
2000	early to mid-Feb.	late July and early Aug.
	early to mid-June	mid- to late Nov.
2001	mid- to late Jan.	early to mid-July
	mid- to late May	mid- to late Sept.
2002	late Apr. and early May	mid- to late Oct.
2003	early to mid-Apr.	late Sept. and early Oct.
2004	late Mar. and early Apr.	early to mid-Sept.
2005	first half of Mar.	late Aug. and early Sept.
		first half of Dec.
2006	last half of Feb.	early to mid-Aug.
	mid- to late June	mid- to late Nov.
2007	late Jan. and early Feb.	
	late May and early June	early to mid-Nov.
2008	early to mid-Jan.	
	early to mid-May	late Oct. and early Nov.
2009	last half of Apr.	first half of Oct.
2010	late Mar. and early Apr.	last half of Sept.

Note: Mercury is visible less than 20 degrees from horizon; always observe from a location with an unobstructed horizon in the specified direction. Data for northern hemisphere only.

VENUS: PERIODS OF PROMINENT VISIBILITY 1996-2010

Western Sky at Dusk	Eastern Sky at Dawn
mid-Oct. 95 to early June 96	late June 96 to late Jan. 97
mid-May 97 to mid-Jan. 98	early Feb. to early Sept. 98
late Dec. 98 to early Aug. 99	late Aug. 99 to mid-Apr. 2000
mid-Sept. 2000 to mid-Mar. 2001	late Apr. to mid-Nov. 2001
early Mar. to late Sept. 2002	mid-Nov. 2002 to mid-Apr. 2003
early Nov. 2003 to late May 2004	late June 2004 to mid-Jan. 2005
early June 2005 to early Jan. 2006	late Jan. to mid-Sept. 2006
mid-Dec. 2006 to mid-July 2007	early Sept. 2007 to late Feb. 2008
late Sept. 2008 to mid-Mar. 2009	early Apr. to mid-Nov. 2009
early Mar. to mid-Sept. 2010	mid-Nov. 2010 to mid-Mar. 2011

Note: Although Venus is the brightest object in the night sky apart from the Moon, it is often close to the horizon. If possible, observe from a location with an unobstructed horizon in the specified direction.

Resources/Photo Credits

The Author

Terence Dickinson is the author of the best-selling guidebook *NightWatch* and is widely regarded as an authority on stargazing techniques for beginners. He is the editor of *SkyNews* magazine, a former editor of *Astronomy* magazine and *The Toronto Star*'s weekly astronomy columnist. He has received numerous national and international awards for his work, among them the New York Academy of Sciences book of the year award and the Royal Canadian Institute's Sandford Fleming Medal for outstanding achievements in communicating science to the public. Asteroid 5272 Dickinson is named after him. He is also an accomplished astrophotographer, having taken most of the pictures in this book himself.

Bernard Clark

Acknowledgments

Although this is my 13th book, it is the first one produced entirely using modern electronic and computer desktop-publishing techniques, and it was quite a revelation for me. I was often in awe, watching over designer Roberta Cooke's shoulder as pages emerged on her Macintosh computer screen as if by magic, appearing essentially as you see them in this printed version. But no magic here. Roberta's graphic skill and attention to detail were the crucial ingredients that made this book possible. Wendy McPeake, publisher of the National Museum of Science and Technology's *SkyNews*, kindly granted permission for use of charts from the magazine, reproduced on pages 20, 22, 48 and 49. Christine Kulyk and Catherine DeLury offered many useful comments. And because Susan Dickinson and Tracy Read of Bookmakers Press tracked every production detail from start to finish, a top-quality product was assured.

Photo Credits

Photographs in this book are by Terence Dickinson except:
p.4 right, Alan Dyer; p.5 top right, Alan Dyer; p.6 left, Space Telescope Science Institute; p.6-7 illustration, John Bianchi; p.7 top, Jack Newton; p.8 top left and bottom left, Alan Dyer; p.11 top, George Liv; p.12 bottom left illustration, John Bianchi; p.16 top left, Frank Hitchens; p.18 illustration, Roberta Cooke; p.20 and p.22, charts from *SkyNews* courtesy National Museum of Science & Technology; p.39 center, Craig McCaw; p.41 top, Jack Newton; p.41 bottom, Jerry Lodriguss; p.45 bottom left, Klaus Brasch; p.45 bottom right illustration, Adolf Schaller; p.46 bottom right, Klaus Brasch; p.46 center left, Jerry Lodriguss; p.47 top right, Klaus Brasch; p.48 and p.49, charts from *SkyNews* courtesy National Museum of Science & Technology; p.51 left illustration, Roberta Cooke; p.51 top illustration, David Egge; p.51 right, Alan Dyer; p.52 top right, Alan Dyer; p.53 top right, Russell Sampson; p.54 lower right, Alan Dyer; p.56 top left, John Hicks; p.56 bottom left, Brian Tkachyk; p.56 top right, Bob Yen; p.57 center and bottom right, Alan Dyer; p.57 top right, Alan Dyer; p.58 top left, Michael Watson; p.58 top right, Richard Keen; p.59 top, Dennis di Cicco; p.59 center, Craig McCaw; p.59 lower right, Andreas Gada.

Resources

As an introductory guide, *Summer Stargazing* is necessarily a brief overview of a vast subject. Here are some of my favorite sources of further information.

ASTRONOMY MAGAZINES

Two major monthly magazines are carried on larger newsstands or are available by subscription: *Sky & Telescope*, Box 9111, Belmont, MA 02178, and *Astronomy*, Kalmbach Publishing, Box 1612, Waukesha, WI 53187. Both are first-rate publications, filled with information on current sky events, discoveries, equipment reviews, and so on.

SkyNews is a colorful bimonthly Canadian astronomy magazine published by the National Museum of Science & Technology. For subscription information and a sample copy (samples to Canadian addresses only), write *SkyNews*, NMST Corp., Box 9724, Stn. T, Ottawa, Ontario K1G 5A3.

OBSERVING GUIDEBOOKS

Immodestly, I'm going to recommend my book *Night-Watch* (Firefly Books, rev. ed., 1993) as the ideal companion volume to *Summer Stargazing*. It covers many topics I didn't have room for in this book, including telescope selection and use, a pronunciation guide, more specific charts and astrophotography techniques and observing tips, to name a few. For even more detail, particularly with regard to equipment, see *The Backyard Astronomer's Guide* (Firefly Books, rev. ed., 1994), which I coauthored with Alan Dyer.

Binocular Astronomy by Craig Crossen and Wil Tirion (Willmann-Bell, Richmond, Virginia, 1992) is the most detailed binocular-observing guide that I have seen; includes an excellent star atlas.

Star-Hopping for Backyard Astronomers by Alan M. MacRobert (Sky Publishing, Cambridge, Massachusetts, 1993) is a well-written introduction to observing with a small telescope.

Turn Left at Orion by Guy Consolmagno and Dan M. Davis (Cambridge, New York, 1989) is another excellent guide for small-telescope observers.

The Guide to Amateur Astronomy by Jack Newton and Philip Teece (Cambridge, New York, 2nd ed., 1995) is a comprehensive handbook for backyard astronomers.

NAKED-EYE ASTRONOMY

Three recommended books that emphasize naked-eye astronomy are: *Sky Phenomena* by Norman Davidson (Lindisfarne Press, Hudson, New York, 1993), *Wonders of the Sky* by Fred Schaaf (Dover, New York, 1983) and *Celestial Delights* by Francis Reddy and Greg Walz-Chojnacki (Celestial Arts, Berkeley, California, 1992).

SKY LORE

Star Tales by Ian Ridpath (Universe Books, New York, 1988) tells the classic Greek and Roman myths associated with each constellation. Myths of other cultures are included in *The New Patterns in the Sky* by Julius D.W. Staal (McDonald & Woodward, Blacksburg, Virginia, 1988).

UNIQUE BOOKS

The Astronomical Companion by Guy Ottewell (Universal Workshop, Greenville, South Carolina, 1979) is a wonderful guide filled with imaginative diagrams that cannot be found elsewhere. Must be ordered direct from the publisher: Universal Workshop, Dept. of Physics, Furman University, Greenville, SC 29613.

Burnham's Celestial Handbook by Robert Burnham (Dover, New York, 1973) is a colossal work in three volumes; a gold mine of information about stars, constellations, nebulas and galaxies.

The Starry Room by Fred Schaaf (Wiley, New York, 1988) covers many aspects of observing the night sky in a series of personal essays containing many useful tips and insights.

ANNUAL GUIDES

Astronomical Calendar by Guy Ottewell (Universal Workshop, Greenville, South Carolina) is a large-format paperback filled with facts and diagrams for the current year. Available from: Universal Workshop, Dept. of Physics, Furman University, Greenville, SC 29613.

Observer's Handbook, published by The Royal Astronomical Society of Canada, contains extensive tables covering virtually every astronomical phenomenon visible from North America. Available from RASC, 136 Dupont Street, Toronto, Ontario M5R 1V2.

ASTRONOMY SOFTWARE, VIDEOS, CD-ROMs

For free catalogs of astronomy slides, videos, posters, computer software, et cetera, contact: Astronomical Society of the Pacific, 390 Ashton Avenue, San Francisco, CA 94112; Hansen Planetarium, 1845 South 300 West #A, Salt Lake City, UT 84115; Sky Publishing, Box 9111, Belmont, MA 02178; The Planetary Society, 65 N. Catalina Avenue, Pasadena, CA 91106.

Index